T0305172

A Research Agenda for Sustainability and Business

Elgar Research Agendas outline the future of research in a given area. Leading scholars are given the space to explore their subject in provocative ways, and map out the potential directions of travel. They are relevant but also visionary.

Forward-looking and innovative, Elgar Research Agendas are an essential resource for PhD students, scholars and anybody who wants to be at the forefront of research.

Titles in the series include:

A Research Agenda for Financial Crime
Edited by Barry Rider

A Research Agenda for Small and Medium-Sized Towns
Edited by Heike Mayer and Michela Lazzeroni

A Research Agenda for Social Welfare Law, Policy and Practice
Edited by Michael Adler

A Research Agenda for Organisation Studies, Feminisms and New Materialisms
Edited by Marta B. Calás and Linda Smircich

A Research Agenda for Terrorism Studies
Edited by Lara A. Frumkin, John F. Morrison and Andrew Silke

A Research Agenda for Sustainability and Business
Edited by Sally V. Russell and Rory W. Padfield

A Research Agenda for Sustainability and Business

Edited by

SALLY V. RUSSELL
> *Professor of Sustainability and Organisational Behaviour,
> Sustainability Research Institute, University of Leeds, UK*

RORY W. PADFIELD
> *Associate Professor of Sustainability and Business, Sustainability
> Research Institute, University of Leeds, UK and Visiting
> Researcher, Universiti Teknologi Malaysia (UTM)*

Elgar Research Agendas

 Edward Elgar
PUBLISHING

Cheltenham, UK • Northampton, MA, USA

Published by
Edward Elgar Publishing Limited
The Lypiatts
15 Lansdown Road
Cheltenham
Glos GL50 2JA
UK

Edward Elgar Publishing, Inc.
William Pratt House
9 Dewey Court
Northampton
Massachusetts 01060
USA

A catalogue record for this book
is available from the British Library

Library of Congress Control Number: 2022950532

This book is available electronically in the **Elgar**online
Business subject collection
http://dx.doi.org/10.4337/9781839107719

ISBN 978 1 83910 770 2 (cased)
ISBN 978 1 83910 771 9 (eBook)
Printed and bound by CPI Group (UK) Ltd, Croydon, CR0 4YY

Contents

Figures

Tables

Contributors

Ralf Barkemeyer is a Professor of Corporate Social Responsibility at KEDGE Business School in Bordeaux, France, and Co-Editor-in-Chief of *Business Ethics, the Environment and Responsibility* (Wiley). Before joining KEDGE, he held positions at Queen's University Belfast Management School and the University of Leeds. Ralf's research focuses on the interface of business, environment and society. Among others, he is interested in the link between CSR and development as well as corporate sustainability performance assessment. Ralf has published in outlets such as *Nature Climate Change, Journal of World Business, Journal of Business Venturing* and *Business Ethics Quarterly.*

Stephen Brammer is Dean and Professor of the School of Management at the University of Bath, UK. Stephen began his academic career at Bath before taking up senior positions at the Universities of Warwick and Birmingham in the UK, and Macquarie University in Sydney, Australia. Stephen is recognised globally for his research in the fields of corporate social responsibility, corporate reputation, stakeholder management, and sustainability. His research focuses on business ethics, corporate social responsibility, and sustainability with an emphasis on firm–stakeholder relationships, the strategic management of these, and corresponding impacts on organisational performance and reputation. Stephen has published in many leading journals, including the *Strategic Management Journal, Academy of Management Perspectives, Journal of Management Studies, Journal of Operations Management, Journal of Organizational Behavior,* and the *British Journal of Management.* Stephen serves as an Associate Editor of the *British Journal of Management.*

Layla Branicki (PhD, Warwick Business School) is a Senior Lecturer at the University of Bath, School of Management. Her interests focus on resilience, crisis management, and corporate social responsibility. Her articles have appeared in such journals as *Academy of Management Discoveries, Academy of Management Perspectives, British Journal of Management, Gender, Work & Organization, The International Journal of Human Resource Management, Journal of Business Ethics,* and *Work, Employment & Society.*

Christian Bretter is a Research Fellow in Radical Transport Decarbonisation at the University of Leeds (UK). He is particularly interested in the intersection of environmental behaviour (such as public transport usage, environmental policy support or food waste), social psychology, and political psychology. Before his current role, he worked as an ESRC-funded Research Fellow in Environmental Psychology with particular focus on better understanding the psychological and behavioural factors driving household food waste. Working with different institutions such as the Waste & Resource Action Programme (WRAP), he has also launched and tested different interventions to reduce household food waste in the UK. After successful completion of this project, he joined the Institute for Transport Studies (ITS) to examine, among other activities, the factors driving transport usage (particularly car use) and, building on this assessment, to explore possible avenues for behaviour change to reduce demand for individual car ownership.

Phani Kumar Chintakayala is a Research Economist at RTI-Health Solutions. He was a Senior Research Fellow at Leeds University Business School and Consumer Data Research Centre at the University of Leeds, UK. Phani's research focused on consumer behaviour and sustainable lifestyles covering consumption patterns of sustainable products. He applied various behaviour modelling techniques to understand consumer preferences in purchasing and using products/services. His areas of research include sustainable consumption, transport, banking, insurance, utilities, and health.

Jo Cutter is a Lecturer in Work and Employment Relations at Leeds University Business School specialising in social-dialogue and the regulation of work notably around training and professional development. Her research is focused on worker voice in climate change mitigation strategies and changing work, jobs and 'green' skills needs and on employer and worker responses to changing labour mobilities, jobs and skills in low-wage sectors. She has led research around the Just Transition including British Academy-funded interdisciplinary research with social science, arts and humanities researchers exploring the future of work in the Just Transition and on worker perceptions of climate change, transition pathways and training needs. She is currently working with a team of 15 researchers from the Global North and Global South studying the US, Canada, the UK, Spain, Germany, Poland, South Africa, Nigeria, Brazil, China and Russia in a comparative study of Just Transition policy and practice.

Alexandra Dales is a Senior Lecturer in Business and Management at York Business School, York St John University, York, UK. Alexandra's background is in economic geography and the interaction of global production networks, market transformation and regulatory/institutional frameworks, particularly in relation to retailing in emerging markets. Alexandra's research has now evolved to examine how sustainability and sustainability transformations are impacted and informed by underpinning market structures and economic geographical processes. Projects include assessing how creative sector market actors communicate the criticality of sustainability through new immersive technologies.

Timothy Devinney is Professor and Chair of International Business at the Alliance Manchester Business School at the University of Manchester (UK). He is also Professor (Conjoint) at the Faculty of Medicine at Macquarie University (Australia). He has published more than a dozen books (e.g., *The Myth of the Ethical Consumer*, Cambridge University Press, with P. Auger and G. Eckhardt, 2010) and more than 100 articles in leading journals including *Management Science*, the *Journal of Business*, *The Academy of Management Review*, *Journal of International Business Studies*, *Organization Science*, *California Management Review*, *Management International Review*, *Journal of Marketing*, *Journal of Management*, *Long Range Planning*, *Journal of Business Ethics* and the *Strategic Management Journal*. In 2008 he was the first recipient in management of an Alexander von Humboldt Forschungspreis and was a Rockefeller Foundation Bellagio Fellow. In 2008 he was elected a Fellow of the Academy of International Business and in 2015 he was elected as a Fellow of the Academy of Social Sciences. He was also appointed as a Fellow of the Royal Society of Arts in 2013. In 2018 he was awarded the Academy of Management's Impact Practice Award for the influence of his scholarship on management practice and policy and in 2019 the Service to the Global Community Award for his work and influence on the global academic community.

Paul Francisco is the Associate Director for Building Science at the Illinois Applied Research Institute at the University of Illinois at Urbana-Champaign, and is the Director of the Indoor Climate Research & Training (ICRT) group. He also holds an appointment as a Senior Research Associate at Colorado State University. His research focuses on the intersection of energy efficiency and indoor air quality with a focus on delivering homes that are both efficient and healthy. His projects are primarily conducted in field environments, and include indoor air quality sampling, energy use logging, and airflow and insulation characterisation of buildings. He is the Director of the ICRT weatherisation training centre which trains home performance practitioners in home evaluation and retrofit.

Zahra Borghei Ghomi is affiliated with the Centre for Corporate Sustainability and Environmental Finance at Macquarie University. Zahra's expertise is in climate-related disclosure and carbon accounting, and her research interest extends to the areas of sustainability, corporate governance and resilience. She has managed and led various research projects in these areas at Macquarie

Business School and external organisations. Zahra has over a decade of experience as a multidisciplinary educator in sustainability, financial accounting, management accounting and introductory finance subjects at Macquarie University to undergraduate and postgraduate students. Zahra worked as an Accounting Practitioner before embarking on an academic career.

Katariina Koistinen holds a PhD (2019) in Environmental Management from the LUT University. Currently she is working as a Postdoctoral Researcher at the University of Turku. Her research focuses on actors in facilitating sustainability change. Her research interests include sustainability transitions, sustainable management and theories of agency. She has authored several publications in journals such as *Energy Research & Social Science, Sustainable Development, Technology Analysis and Strategic Management,* and *Journal of Cleaner Production.*

Dunja Krause is a Research Officer at the United Nations Research Institute for Social Development (UNRISD) where she leads the Environmental and Climate Justice Programme. She co-coordinates the Just Transition Research Collaborative and co-authored the reports 'Mapping Just Transition(s) to a Low-Carbon World' (2018) and 'Climate Justice from Below. Local Struggles for Just Transitions(s)' (2019). Dunja also co-edited the volume *Just Transitions: Social Justice in the Shift Towards a Low-Carbon World* (Pluto Press, with Edouard Morena and Dimitris Stevis, 2019). A geographer by training, she has previously worked at the United Nations University Institute for Environment and Human Security, focusing on vulnerability assessments to natural hazards, the development of a global risk index, and the evaluation of climate change adaptation options.

Martina K. Linnenluecke is Professor of Environmental Finance and leads the Center for Corporate Sustainability and Environmental Finance at Macquarie University, Sydney, Australia. Martina's research focuses on the strategic and financial implications of corporate adaptation and resilience to climate change impacts. She has published over 100 academic articles, book chapters and conference papers and has been the recipient of numerous awards for her work, including the Carolyn Dexter Best International Paper Award at the Academy of Management Conference. She is the author of the book *The Climate Resilient Organization* (Edward Elgar Publishing, 2015) and has extensive experience in working with government and industry on climate adaptation strategies. She is currently the Program Chair of the ONE Division of the Academy of Management.

Ning Lu is an Adjunct Faculty at Leeds University Business School, Visiting Fellow at the School of Psychology in the University of Leeds and Senior

Research Fellow at Bradford Institute of Health Research. He earned his ESRC-funded PhD at the University of Leeds and Master's Degree at Imperial College London. His academic research interest is focused on consumer behaviour, personality traits and data science. He is also the co-founder of Leeds Data Science Society and led the University of Leeds to win the national championship in the Hiscox Data Challenge in 2017. In 2018, he was nominated as 'One to Watch' by Leeds Digital Festival.

Marileena Mäkelä is a Senior Lecturer of Corporate Environmental Management (CEM) in the Jyväskylä University School of Business and Economics. She holds a doctoral degree from Turku School of Economics since 2020. She has a Master's Degree in Engineering from Tampere University of Technology and a Master's Degree in Economics from the University of Jyväskylä. Her research interests relate to sustainability reporting, employees' role in sustainability, the circular economy and futures research. She teaches in the international CEM Master's programme, in Turku School of Economics and in Tampere University. She has published her research in journals such as *Journal of Cleaner Production, Forest Policy and Economics, Business Strategy and the Environment, Sustainable Development* and *Futures.*

Suzana Matoh is a Lecturer in the Department of Strategy, Enterprise and Sustainability at the Manchester Metropolitan University Business School. She is a programme leader for MSc Management and Sustainability. Her research interests focus broadly on drivers and the process of organisational change for sustainability. She is interested in factors that influence changes in business models to achieve improved sustainability performance, for instance, material demand reduction. She uses cognitive mapping technique to explore sustainability issues that drive managerial decision making for business model change. The research projects she was involved with focused on a range of sustainability-related topics, from exploring how trade unions respond to climate change issues and green economy to how different actors in the palm oil supply chain address traceability and transparency.

Jyoti Mishra is an Associate Professor in Information Management at Leeds University Business School, University of Leeds, Leeds, UK and currently serves as a Deputy Director of AIMTech Research Centre. Jyoti comes from an Electronics Engineering background and is interested in investigating how technology can be used for sustainability and circularity. Her research interests are in the circular economy, technology use in organisations, information management, and sustainable supply chains. Before joining Leeds, Jyoti worked as a Lecturer in Sustainable Operations Management. She has taught the world's first MBA in Innovation and Circular Economy at the University of Bradford. Jyoti has published in several international top-ranked journals on the topic of Circular Economy, Supply Chain, and Information Management. She is an Associate Editor for the *European Journal of Information Systems* and Subject Editor at the *Sustainability* journal. She acts as a peer reviewer for several international journals.

Iana Nesterova is a Postdoctoral Fellow at the Department of Geography, Umeå University, Sweden. Her PhD focused on small business transition towards degrowth. Her research on business and degrowth transformations has been published in the leading journals in the field of sustainable organisations, including *Ecological Economics, Journal of Cleaner Production* and *Futures*.

Alesia Ofori is a Research Fellow in Water and Sanitation Governance at the School of Politics and International Studies at the University of Leeds. Her research is broadly in the politics and anthropology of resource governance and development in sub-Saharan Africa, focusing on the extractive industries, water, sanitation, and agriculture. She holds a PhD in International Development from the University of Leeds. Her research into natural resource extraction and management incorporates theories and concepts from politics, anthropology, history and ecology to showcase the "more than technical" nature of resource governance issues. Her PhD thesis which investigated the political ecology of the water-mining nexus in Ghana showcased the political complexities associated with preventing artisanal miners from polluting surface water bodies. Her research outputs have been published widely in Journals such as *World Development, One Earth* and *Extractive Industries and Society*.

Tiina Onkila holds a PhD in Corporate Environmental Management at the University of Jyväskylä (2009), and was Adjunct Professor at the University of Turku (2018). She currently works for the University of Jyväskylä, and is responsible for the Corporate Environmental Management Master's programme. Her research focuses on sustainable business, stakeholder relations, the employee–sustainability relationship and sustainability agency. Her research has been published in journals such as *Organization & Environment, Journal of Business Ethics, Business Strategy and the Environment* and *Journal of Cleaner Production*. She is currently engaged in the CICAT2025 research project with a focus on agency for a sustainable circular economy and the BIODIFUL research project with a focus on biodiversity respectful leadership.

Alice Owen is Professor in Business, Sustainability & Stakeholder Engagement in the Sustainability Research Institute at the University of Leeds, UK. She uses her background in policy development and industry to guide her teaching, helping postgraduate students of sustainability connect theory with practical application and change. Her research on the construction industry has focused on the realities of mainstream construction practice carried out by sole traders and small firms.

Rory W. Padfield is an Associate Professor in Sustainability and Business in the Sustainability Research Institute at the University of Leeds, following academic positions at Oxford Brookes University (2016-2019) and Universiti Teknologi Malaysia (2010-2016). In his teaching and research, Rory engages with the broad themes of uneven development in the Global South, natural resource governance, business practices and sustainable management of supply chains, and media and digital geographies. Rory draws on a variety of academic perspectives to pursue critical questions around ethics, social and environmental responsibility, and local–global political economy. He has published his research in various academic journals, including *Business Strategy and the Environment, Environment and Planning E, Transactions of the Institute of British Geographers, Frontiers, Global Change Biology, Landscape Research, Clean Technologies and Environmental Policy, Waste Management* and *Journal of Cleaner Production.*

Effie Papargyropoulou is an Associate Professor in Sustainable Food Networks at the University of Leeds. Effie is interested in the interactions between people and the environment, and the management of the impacts human activity has on the environment. She has explored these interactions in the context of global environmental challenges such as climate change and food security, with a focus on sustainability and sustainable development. Effie's research focuses on two main themes: food systems, food waste management and food consumption; and low carbon cities, decarbonisation of human activities and climate change mitigation. Before joining the University of Leeds, Effie worked as a lecturer in Malaysia for six years, and before that as an environmental consultant in the UK.

Rebecca Pieniazek is an early career Lecturer in Organisational Behaviour at Leeds University Business School and a member of the Workplace Behaviour Research Centre and the Socio-Technical Research Centre. She is interested in understanding and helping improve both the personal resilience of employees and the resilience of organisations. Alongside this she is passionate about examining and improving employees' motivation, and how they can successfully juggle multiple and often conflicting goals in the workplace. Rebecca is an expert in teaching performance management and career development, as well as research methods.

Jonatan Pinkse is a Professor of Strategy, Innovation, and Entrepreneurship at and Executive Director of the Manchester Institute of Innovation Research

(MIoIR), Alliance Manchester Business School, the University of Manchester. His research interests focus on corporate sustainability, business model innovation, social entrepreneurship, cross-sector partnerships and the sharing economy. In his research, Jonatan analyses how firms make strategic decisions to adapt to a more sustainable economy and deal with the ensuing tensions between issues and actors. He also investigates barriers to firm adoption of disruptive technologies from cognitive, organisational and institutional perspectives. Before moving to Manchester, he held positions at the Universiteit van Amsterdam and Grenoble Ecole de Management. Jonatan has authored more than 50 scholarly and practitioner articles in journals such as the *Academy of Management Review, Journal of Management Studies, Research Policy, Journal of International Business Studies, Organization Studies, California Management Review, Journal of Business Venturing* and *Entrepreneurship Theory and Practice.* He is associate editor at *Organization & Environment* and *Business & Society* and past chair of the ONE Division of the Academy of Management.

Samuel Howard Quartey is a Lecturer in Human Resource Development at the Department of Adult Education and HR Studies. He serves as Head, at the University of Ghana Learning Centre in Koforidua. Before joining the University of Ghana, he held other academic positions at Central University, Ghana, as a Senior Lecturer; University of Adelaide, Australia as a Casual Lecturer; and Webster University, USA, as an Adjunct International Faculty. Sam holds a PhD in Management (with specialisation in Knowledge and Sustainable Development) from the University of Adelaide, Australia. He also holds a Bachelor's Degree in Psychology; and a Master of Philosophy degree in Human Resource Management from the University of Ghana. Sam also serves as an External Assessor for Ghana Tertiary Education Commission (GTEC). He also serves as External Examiner for private and public universities in Ghana. He is an Editorial Board Member for the *Business Strategy and Development Journal,* Wiley, as well as a Reviewer for international peer-reviewed high impact quality journals. His current research examines critical issues in Human Resource Development including competence, training, learning, development, education, human capital, health and safety, knowledge, leadership, and sustainability. He is a member of the Academy of Human Resource Development, USA and Development Studies Association, UK.

Ben Robra is a Postdoctoral Researcher at the Universidade de Vigo's Post-growth Innovation Lab. His research focuses on economic organisations and innovation in connection to degrowth as well as postgrowth. Ben holds a BA in Business Administration from the Hamburg School of Business

Administration as well as an MSc in Ecological Economics and PhD in Sustainability from the University of Leeds.

Katy Roelich is an Associate Professor and Co-Director of the Sustainability Research Institute in the School of Earth and Environment at the University of Leeds. Prior to joining academia Katy had nine years' experience in environmental and engineering consulting in the UK and overseas. Her current research activities centre on long-term decision making under uncertainty, particularly in complex socio-technical systems where multiple actors must interact to make decisions in the face of deep uncertainty. She also researches public participation in transformation of complex systems. She has developed tools, indicators, participatory models and decisions support approaches to help decision makers understand this uncertainty and improve decision making.

Sally V. Russell is Professor in the Sustainability Research Institute at the University of Leeds, UK. Her research focuses broadly on behaviour change for sustainability across domains including water, energy, food waste, and workplace pro-environmental behaviour. Her research also examines how emotional reactions to environmental issues affect subsequent behaviour and decision making – both within and outside organisations. Her research attracts funding sources including industrial partners, European Commission, British Academy, National Climate Change Adaptation Research Facility (Australia) and local and state governments (UK and Australia). Her work has been published in journals including: *Resources, Conservation and Recycling*; *Water Resources Research*; the *Journal of Environmental Management*; and *Business Strategy and the Environment*. She currently serves on the editorial boards of *Business, Strategy and Environment*, the *Journal of Organizational Behaviour*, and the *Journal of Management & Organization*.

Laura Smith is a Lecturer in Sustainability and Business in the Sustainability Research Institute at the University of Leeds. Her research interests are broadly in the extractive industries and social justice, corporate responsibility and international development, and the politics of justice and inclusion in the low carbon transition. Laura teaches on the online Sustainable Business Leadership MSc at the University of Leeds. Prior to pursuing her PhD at the Sustainability Research Institute in 2012, Laura worked in the third sector for ten years, including for an ethical trading NGO supporting smallholder farmers, and as a human rights campaigner.

Thomas Smith is an Associate Professor in Environmental Geography at the London School of Economics and Political Science. He teaches on a number of environmental courses, focusing on innovative technology-enhanced experiential learning and field-based education in geography. Tom is a geographer and environmental scientist, specialising in interdisciplinary approaches to understanding the role of biomass burning in the Earth system. Tom enjoys highly collaborative research focusing on greenhouse gas and reactive emissions from wildland fires in savannas and tropical peatlands. He is particularly interested in complex interactions between agricultural practices, land degradation, fire emissions characteristics and their associated impacts. His expertise includes infrared and VNIR spectroscopy, tropical environmental change, wildfire spread modelling, knowledge exchange, and land management decision support.

Kari Solomon is a PhD student at the University of Leeds in the School of Earth & Environment, under the Sustainability Research Institute (SRI), with her research project serving the objectives of Businesses & Organisations for Sustainable Societies (BOSS). Kari holds an academic and practitioner perspective, having earned a Bachelor's in Management at Hiram College, a Master's of Business Administration (MBA) in Sustainability at Baldwin Wallace University, and earned 13 years of work experience in the public transportation industry with responsibilities in financial management, strategic planning, performance excellence, and sustainability strategy and reporting. Kari's current research project at the University of Leeds seeks to understand the links between corporate sustainability management practice and impacts on society, measured by sustainability performance and quality of life indicators.

Satu Teerikangas is Professor of Management & Organization, University of Turku (UTU), Finland, and Honorary Professor in Management, University College London, UK, where she was tenured (2010-15). With 150+ publications, 2M€+ research funding, her work on mergers & acquisitions is recognised (BBC, Forbes). She is editor of the *Handbook of M&As* (Oxford University Press, 2012) and the *Research Handbook on Sustainability Agency* (Edward Elgar Publishing, 2021). She currently leads the Strategic Research Council-funded 'CICAT2025 Transitions to Circular Economy' project's work package on agency and is a vice leader for the 'BIODIFUL Biodiversity respectful leadership' consortium. She was UTU's Executive Educator of the year (2020). Prior to an academic career, Satu worked in the oil and gas industry in the Netherlands and the UK.

Vera Trappmann is a Professor in Comparative Employment Relations at Leeds University Business School. Her research examines labour relations across Europe. Her main research interests focus on the dynamics of organisational restructuring and its impact on working biographies, and organised labour. Drawing on her work on transitions from state-planned economies to capitalism, she studies climate change and decarbonisation and its impact on work and employment. She has published more than 60 monographs, articles, chapters and reports. Her work has been funded by the European Commission, the German Research Foundation, Hans-Boeckler Foundation, Otto-Brenner Foundation, Polish–German Science Foundation, German Academic Exchange Service, German Ministry of Research and Education, Federal states and trade unions. Currently she is the PI of a project on Just Transition policies and practices in 11 countries.

Kerrie L. Unsworth is Chair of Organisational Behaviour at Leeds University Business School and a member of the Workplace Behaviour Research Centre. She is interested in understanding and helping people at work, including their pro-environmental behaviours, creativity, well-being and productivity; she is an expert in how people juggle their different goals and identities and the effect that has on their motivation and behaviours. Kerrie has over 25 years' experience teaching concepts such as well-being and leadership to executives and students and is a passionate believer in helping to create a better world. She has published in a range of top academic journals including *Academy of Management Review* and *Journal of Applied Psychology*, she is an Associate Editor at *Human Relations* and she sits on the Editorial Boards of *Journal of Applied Psychology*, *Journal of Management Studies* and *Journal of Occupational & Organizational Psychology*.

James Van Alstine is Associate Professor in Environmental Policy at the Sustainability Research Institute at the University of Leeds. His research is focused on the political ecology of resource extraction, the politics and governance of energy transitions, and opportunities for low carbon, inclusive and climate resilient development. James seeks to bridge the academic-practitioner divide by pursuing action-oriented research that aims to maximise policy and pro-poor development impacts. He works closely with development partners, policy makers and regulatory agencies at the international, national and local levels, as well as with industry, NGOs and communities. James' recent research explores social acceptance of hydrogen technologies.

Chee Yew Wong is a Professor of Supply Chain Management at Leeds University Business School, University of Leeds, UK. He has more than nine years of industrial working and consultancy experience in operations, purchasing, production, inventory and distribution management and supply chain design with SMEs and multinational companies specialised in beverage, retail, consumer goods, toys, engineering, metal production, and polymer distribution. His research interests lie in the areas of supply chain integration, digital supply chain, supply chain analytics, green supply chain. He is a co-author for a book called *Sustainable Logistics and Supply Chain Management* (Kogan

Page, 2013). He is the editor-in-chief of the *International Journal of Physical Distribution and Logistics Management*.

William Young is Professor and Chair of Sustainability and Business at the Sustainability Research Institute at the University of Leeds, UK. William's research is focused on consumer behaviour around sustainability issues working with retailers, suppliers and consumer organisations. He develops theoretical frameworks and applied tools that understand and change consumer behaviour in purchasing and using products. The frameworks and tools use indicators to show measurable reductions in environmental impacts. William is Associate Editor of the *Resources, Conservation and Recycling* journal and on the journal Editorial Boards of *Corporate Social Responsibility & Environmental Management* and *Social Business*. William is co-organiser of the annual Corporate Responsibility Research Conference along with Kedge Business School in France.

Qinghua Zhu is a Distinguished Professor in Antai College of Economics & Management and Associate Dean of Sino–US Global Logistics Institute at Shanghai Jiao Tong University. Her main interests are sustainable supply chain management. She serves as an editor of *International Journal of Production Economics* and an associate editor of *International Journal of Operations Management*, and has been a special issue editor for several journals such as *European Journal of Operational Research* and *IEEE Transaction on Engineering Management*. She is currently an editorial or advisory board member for ten English journals, including *Business Strategy and the Environment*.

Abbreviations

AFP	Analytical Fingerprint
AI	Artificial Intelligence
ASM	Artisanal and small-scale mining
BIA	Benefit Impact Assessment
CBDR	Common but Differentiated Responsibility
CDSB	Climate Disclosure Standards Board
CE	Circular Economy
COPs	United Nations Climate Change Conferences
CPLC	Carbon Pricing Leadership Coalition
CRM	Critical Resource Materials
CSR	Corporate Social Responsibility
DRC	Democratic Republic of the Congo
EITI	Extractive Industries Transparency Initiative
EKC	Kusnets Curve
ESG	Environment, Social and Governance
EVs	Electric Vehicles
FDI	Foreign Direct Investment
FMCG	Fast-Moving Consumer Goods
FPIC	Free, Prior and Informed Consent
FSB	Financial Standards Board
GDP	Gross Domestic Product
GEC	Global Environment Centre
GHG	Greenhouse Gas
GIS	Geographic Information Systems

GRI	Global Reporting Initiative
GSCM	Green Supply Chain Management
IAQ	Indoor Air Quality
ICE	Internal Combustion Engine
ICT	Information & Communication Technology
IFC	International Finance Corporation
IFRS	International Financial Reporting Standards
ILO	International Labour Organization
IoT	Internet of Things
IPCC	Intergovenmental Panel on Climate Change
IPE	Institute for Public & Environmental Affairs
ISAE	International Standards on Assurance Engagement
M&As	Mergers and Acquisitions
M&S	Marks & Spencer
MMSD	Mining Minerals and Sustainable Development
NGO	Non-governmental Organisations
OCAW	Oil, Chemical and Atomic Workers Union
OCMAL	Observatory of Mining Conflicts of Latin America
OECD	Organisation for Economic Co-operation and Development
OES	Organisational Environmental Sustainability
OESR	Organisational Environmental Sustainability and Resilience
OR	Organisational Resilience
PH	Porter Hypothesis
PV	Photovoltaic
RMI	Repair, Maintenance and Improvement
SAPP	Strategy as Practice and Process
SASB	Sustainability Accounting Standards Board
SCM	Supply Chain Management
SDGs	Sustainable Development Goals
SEE	Society of Entrepreneurs & Ecology
SMEs	Small and medium-sized enterprises

TCFD	Task-force for Climate-related Financial Disclosure
TKG	Transformationskurzarbeitergeld
TPB	Theory of Planned Behaviour
TRA	Theory of Reasoned Action
TUC	Trades Union Congress
UAS	Unmanned Aerial Systems
UK CCC	UK's Committee on Climate Change
UNGC	United Nations Global Compact
USDOE	US Department of Energy
USEPA	US Environmental Protection Agency
USHUD	US Department of Housing and Urban Development
VRF	Value Reporting Foundation
VtoX	Vehicle-to-Everything
WoS	Web of Science
Y&H	Yorkshire and Humber

1 An introduction to *A Research Agenda for Sustainability and Business*

Sally V. Russell, Rory W. Padfield and Christian Bretter

The role of business in advancing sustainable development is no longer debated. But how businesses can and are acting to redress social and environmental issues is an area of growing academic and practical interest. The rise of business has been responsible for significant advances since the Industrial Revolution, but businesses have also been responsible for contributing to many of the problems we face today. There is a long history of research that highlights the threats to and collapse of the planet's natural ecology and ecological systems. These include a changing climate, resource depletion and degradation, pollution, biodiversity loss and irreversible impacts to life-supporting ecosystems. Research has also stressed the impacts of business on social issues, such as uneven development, food insecurity, global inequality and rising poverty.

Despite the known impacts of business on social and environmental systems, there is also a growing sense that businesses make up an important part of the future solution. Businesses have been recognised as one of several critical stakeholders in the achievement of international frameworks for global sustainability (e.g., the United Nations Sustainable Development Goals), and many sectors and industries are regarded as pioneers in technological innovation for advancing sustainability policy and practice. Notwithstanding the evolving constraints and challenges facing contemporary businesses – whether small, medium and large – they are ideally placed to contribute in a genuine and meaningful way to global sustainability.

Within this book we bring together a range of interdisciplinary perspectives to identify and highlight the need for a whole of systems approach to sustainability and business. We note that research within the field of sustainability and business is often dominated by a narrow focus with single levels of analysis and with a single dominant perspective. By bringing together the collection of chapters in this book, we aim to highlight the breadth and depth of levels of analyses and perspectives that, we argue, are needed to set a future agenda for sustainability and business research and to accelerate actions that address the challenges of sustainability at a whole of system level. The book is structured according to the dominant level of analysis of each chapter, moving from the individual, to organisational, to system levels, with interrelationships between levels highlighted at each stage (Starik & Rands, 1995). We also note how our chapter authors cover a broad range of perspectives including but not limited to instrumental, economic, network, and political perspectives (Bondy & Matten, 2012).

We have designed the book to provide a valuable resource for scholars new to the area, and established researchers looking to develop or deepen their focus on sustainability and business research. Thus, it is our hope that this book will enhance the reader's understanding of the history, current state, and future questions in sustainability and business research. The aims of the book are threefold: First, this book aims to evaluate the current 'state of the art' of sustainability and business as a field. To do this each chapter gives a background or history of the development of a particular line of research questioning and provides an overview of where current research has led us. The second aim is to discuss the key challenges for the field and its contribution to assessing progress to a more sustainable practice. This interrelationship between research and practice is particularly important in the context of sustainability and business given the extent to which social science research has an influence on practice and policy. The final aim of this book, and indeed the aim of each chapter, is to provide a future research agenda for new and existing researchers by illustrating the path forward for producing research that has an impact on the science of business and sustainability and also a positive impact on practice.

Throughout this chapter and the remainder of the book, we predominantly refer to the relationship between 'business' and 'sustainability'. We use these terms broadly and do not confine our definition of business to one type of profit-making firm or organisation. Rather we use the term 'business' to capture a broad spectrum of businesses and organisations, whether they be for-profit, not-for-profit, social enterprises or corporations. Thus, the term business should be interpreted within the widest possible perspective. Similarly, we use the term 'sustainability' in its broadest sense to capture the range of social, environmental, and economic issues facing businesses today. We align our definition of sustainability to that of sustainable development as "meeting the needs of the present without compromising the ability of future generations to meet their own needs" (Brundtland, 1987, p. 37), yet we do not constrain our definition to a primary focus. Each of our chapter authors prioritise social and/or environmental issues slightly differently and we have intentionally presented these divergent perspectives to capture the breadth of understanding of the term 'sustainability'. We leave it to each of our authors to define and use the most relevant terminology to their primary areas of expertise and thus you may notice some variation in how key terms are defined and described.

In the following section, we begin by introducing a broad history of the development of sustainability and business as a unique and legitimate research field. We then introduce four broad perspectives, or understandings, that dominate the literature in the field of sustainability, ranging in a spectrum from instrumental to value-driven perspectives (Bondy & Matten, 2012). Following this we outline three broad levels of analysis and briefly examine differing perspectives at each level (Hahn, Pinkse, Preuss, & Figge, 2015; Starik & Rands, 1995; Whiteman, Walker, & Perego, 2013). The final section of this chapter sets the stage for a future research agenda of sustainability and business by giving an outline of each of the chapters and situating them within their dominant level/s of analysis and dominant perspective/s. By doing so we aim to facilitate the essential work of future researchers who will progress both research and practice in this important area and who will bring the field forward into new and exciting directions; all with the aim of improving the sustainability of business and in so doing improving the future sustainability and flourishing of all living beings.

A history of sustainability and business

The relationship between 'sustainability' and 'business' has its roots in Corporate Social Responsibility (CSR) and the idea that businesses could (or should) have a concern for broader society over and above what is needed to generate profit (Carroll, 1999). It was, however, not until the mid-20th century when the idea of CSR started to gain more traction in the literature (Bowen, 1953; see also Kreps, 1940), which in turn gave rise to the view that businesses should and could act responsibly and sustainably (for a review, see Carroll, 1999). In the following paragraphs, we elaborate more on the historical details of this amalgamation between business and sustainability.

Starting from the 1960s, we show that the attention to the concept of a firm's environmental or social sustainability has had ebbs and flows, partly driven by environmental disasters and subsequent legislation. Hoffman and Bansal (2012) refer to this progression in terms of three waves of corporate environmentalism. They eloquently elucidate how shifts in public understanding of businesses environmentalism resulted in significant changes to the amount of attention the firm paid to its sustainability performance.

The first wave occurred in the late 1960s and it was enabled by Rachel Carson's (1963) milestone publication, *Silent Spring*. Carson's powerful account brought awareness to the destructive impact of chemicals on nature and sparked a plethora of important events and initiatives related to the environment. These included the creation of the Club of Rome in 1968, a committee of scientists and economists to better understand industry's impact on the environment, and the first global conference to focus on environment issues, the Human Environment Conference organised by the UN General Assembly among others in 1972 (Hoffman & Bansal, 2012). These events highlighted the impact of business on the natural environment, which resulted in the establishment of regulations and the setting of new environmental standards for various industries (Brenton, 1994; Jones, 2017). Despite expectations that corporations would comply with such regulations, efforts remained low and were largely focused on compliance with the law (Hoffman, 2001). In short, environmental regulations were often viewed as restrictive to economic growth (Carroll, Lipartito, Post, & Werhane, 2012; Hoffman & Bansal, 2012).

A second wave of corporate environmentalism or sustainability followed in the late 1980s and early 1990s (Hoffman & Bansal, 2012). In this wave, corporations drastically changed their approach to environmentalism and moved from a compliance exercise to one that was considered a strategic matter. At least in part, the reasons for this shift were major environmental disasters that sparked fear amongst the public alongside distrust towards businesses. Among those was an accident in 1984 where a gas leakage from a pesticide plant in Bhopal, India led to the death of 3,500 people whilst wounding a further 300,000. There were also several other events that raised the public's attention to corporations' (un)sustainability, such as the Chernobyl nuclear disaster in 1986 and the release of the Brundtland Commission report (1987, p. 37), calling for sustainable development that would meet "the needs of the present without compromising the ability of future generations to meet their own needs". At the same time, there was evidence of growing public environmental consciousness, with individuals seeking 'greener' products. Due to the growing pressure and the subsequent focus on the firm's environmental impact, companies understood sustainability as a strategic opportunity that could add value and generate profit (Bergquist, 2019; Hoffman & Bansal, 2012). Thus, businesses started to incorporate environmentalism into their business strategy (Hoffman & Bansal, 2012).

The third wave, which is arguably still ongoing, commenced around the year 2010 and amalgamates not only environmental, but also social concerns on a global scale (Hoffman & Bansal, 2012). The public's rising awareness of the myriad of issues that expand beyond climate change to include national security and religious morality, has pressured companies to incorporate sustainability-focused actions into their business strategy. Accordingly, this wave of public attention has led businesses to address sustainability as more than just environmental protection, but rather as a form of responsible and ethical business practice (for a more detailed review see Bergman, Bergman, & Berger, 2017).

Despite the rise of sustainability within business strategy, many of the problems identified in the 1960s remain and indeed, continue to worsen (IPCC, 2022; Whiteman et al., 2013). The Planetary Boundaries Framework, for example, identifies just how many of the Earth's critical systems are at or beyond a tipping point (Whiteman et al., 2013). As Hoffman and Ehrenfeld (2015) point out, this leads to the inevitable conclusion that the third wave of sustainability and business has not addressed the root causes of environmental and social problems facing the world today. In making this argument, Hoffman and Ehrenfeld (2015, p. 228) argue that we may have entered a fourth wave, that of the Anthropocene; "a new geologic epoch in which we cannot talk about the Earth's ecosystems without recognizing the human role in altering them". This fourth wave requires a fundamental shift in the understanding of the relationship between humans and nature with consideration of scientific, social, economic, and ethical dimensions (Oldfield et al., 2014).

Whiteman et al. (2013) argue that there is a need to move towards a whole of systems approach, one that is increasingly being argued for within both management and sustainability literature (e.g., Whiteman et al., 2013). Within this book, many of the future research questions proposed align with this fourth wave in a departure from standard management theory and conventional approaches. We integrate sustainability with business in a way that aims to lead you, the reader, to think deeply about what is required in the field of sustainability and business research to create meaningful change for our species and our planet.

The development of research on sustainability and business

Despite increased public attention towards a firm's sustainability during the 1990s, scientific research in the area in the field of 'sustainability and business' did not accelerate until around 2010 onwards, echoing the third wave of corporate environmentalism (Hoffman & Bansal, 2012; Hoffman & Ehrenfeld, 2015). A keyword search of the scientific database Web of Science (WoS) with the term 'Sustainability and Business' from 1970 onwards shows the growing interest and development of the field over time (see Figure 1.1).

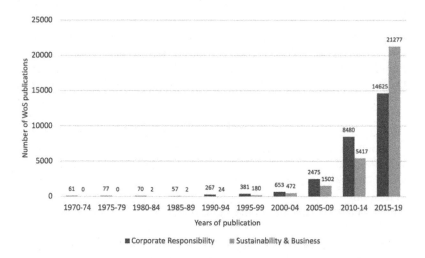

Figure 1.1   Number of WoS publications for the terms 'Sustainability and Business' and 'Corporate Responsibility', 1970–2019

In light of earlier discussions in this chapter concerning the history of sustainability, an attentive reader might wonder why there were not many articles before the 1980s. One reason for this seemingly late awakening of the scientific community, and particularly in management studies, may relate to the terminology used in our article search. Recall that the term sustainability (or sustainable development) became popular in 1987 with the publication of the UN's Brundtland Commission report (Brundtland, 1987). Before 1987, sustainability was covered under the term 'Corporate Responsibility', stemming from a generic sense of a firm's responsibility towards society (Carroll, 1999). Indeed, a search in the WoS with the term Corporate Sustainability shows literature as early as 1907. Although, the number of publications grew considerably from around the year 2010 as illustrated in Figure 1.1, we can also see publications in the 1970s and the 1980s. In the five years to 2019, the figure shows that publications on the topic of 'Sustainability and Business' are more frequent than those of 'Corporate Responsibility', thus demonstrating the maturity, relevance, and academic focus on this area of research.

Sustainability and business across different perspectives

There are many different understandings of how to approach sustainability from the perspective of business. The broad spectrum of approaches can be categorised along a continuum from instrumental approaches at one end, and value-focused approaches on the other (Bondy & Matten, 2012; Windsor, 2006). Using this continuum, Bondy and Matten (2012) outline four different approaches: instrumental, economic, network, and political. We briefly outline each below.

Instrumental approaches to sustainability and business can be understood as considering the biosphere as a limited resource, which can, and should, be used as a source of gaining competitive advantage (McWilliams & Siegel, 2001). Although proponents of this view advocate for environmental protection, the underlying motivation for sustainability action for businesses from this perspective is to create competitive advantage (Bondy & Matten, 2012; Hart, 1995; Kurucz, Colbert, & Wheeler, 2008). The instrumental approach to sustainability emphasises the development of competitive advantage through strategies such as cost reduction (Hart, 1995), improved reputation (Fombrun, 2005) or value added in the creation of 'green' products (Shaw & Clarke, 1999). Meta-analytic evidence supports a somewhat weak link between sustainable actions and financial performance (Margolis, Elfenbein, & Walsh, 2009; Wu, 2006), yet instrumental approaches assume that sustainability actions will result in positive financial impacts (Bondy & Matten, 2012). This understanding of sustainability and business has elicited a plethora of academic debate, particularly given that it largely prioritises the liberty of the firm over and above responsibility towards the environment (Banerjee, 2012).

In contrast, economic approaches to sustainability and business are more focused on pricing the use of environmental and social resources (Bondy & Matten, 2012; Daly, 1998). Just as economics consider financial, physical, and human capital when estimating the price of production, the economic perspective rests on the assumption that the price of production to the environment and society should also be considered (Costanza et al., 1997). The reasoning behind this view is that once the price of a good or service includes the costs to the environment, an economic incentive exists to minimise this cost, thereby contributing to environmental protection (Bondy & Matten, 2012; Cormier, Magnan, & Van Velthoven, 2005). While both instrumental and economic approaches seem to rely on monetary reasoning to justify sustainability-related actions of the firm, the underlying motivation or assumptions differ. While instrumental approaches are based on maximising financial benefit to the firm, economic approaches use the monetarisation of the environment and society in order to protect and sustain these domains (Bondy & Matten, 2012). From both instrumental and economic perspectives, the view of business is largely passive. Conversely, alternative perspectives suggest a much more active role for business.

Network approaches view firms as profoundly embedded in both society and the environment and businesses are understood to have a responsibility towards the environment (Bondy & Matten, 2012). Accordingly, network approaches are more towards the value-based end of the continuum as compared to more instrumental perspectives (Windsor, 2006). Although there are a range of different network approaches (see Bondy & Matten, 2012 for more detailed discussion), we focus here on stakeholder theory (Freeman, 1984). Stakeholder theory posits that corporations not only have essential relationships dependent on employees, suppliers, or customers to effectively manage their business, but also that the government and the environment should be understood as important stakeholders as well (Chang et al., 2017; Freeman, 1984; Lozano, Carpenter, & Huisingh, 2015). Given that organisations ultimately aim to satisfy their stakeholders, such stakeholders can put pressure on the organisation, thereby influencing its actions (Linnenluecke & Griffiths, 2013). Having a myriad of stakeholders, however, also comes with an increased complexity. For example, organisations may often receive conflicting demands from multiple stakeholders (Kassinis, 2012).

Finally, political approaches deal with the grey area of whose responsibility environmental protection should be (Bondy & Matten, 2012). Traditionally, government was seen as having responsibility to act in a way beneficial to society and the environment (Moon, Crane, & Matten, 2005). When corporations are expected to take on some of that responsibility, and to be held accountable when standards were not being met, they too become political actors (Crane, Matten, & Spence, 2008; Scherer, Palazzo, & Matten, 2009). There are a number of theories that take this perspective. For example, new corporate environmentalism (Jermier, Forbes, Benn, & Orsato, 2006) refers to a set of ideas that act as a management guide outlining the need for organisations to proactively control their impact on the environment above and beyond environmental legislation and regulation (Bondy & Matten, 2012). This theory suggests that businesses act within the political domain by shaping the public discourse on environmental protection and social issues (Forbes & Jermier, 2012; Jermier et al., 2006). Given that the theoretical perspective of new corporate environmentalism places value on the environment and suggests that business should proactively protect it, the approach stands in contrast to more instrumental approaches that argue for a protection for financial benefit.

Our aim in this brief review was to highlight the broad spectrum of perspectives in understanding the relationship between sustainability and business, rather than provide an exhaustive list of perspectives. Our review shows that some perspectives, such as instrumental and economic, assume businesses as more passive actors, others, such as the network and political approaches, argue for a much more active role and accountability for wider progress on sustainability issues. As you, the reader, will see throughout this book we have not constrained our authors to one perspective or another. Rather, the variety of the chapters presented here serve to highlight the diversity of perspectives in sustainability and business research which will create awareness to the broad spectrum of current research. In turn, we hope readers will recognise the value of integrating differing perspectives in order to progress the field into the future.

Sustainability and business across different levels of analysis

Just as there are differing perspectives taken to examine the relationship between sustainability and business, so too are different foci found in levels of analysis. Much of the early research in sustainability and business was focused on the level of the firm as the unit of analysis (Hoffman & Bansal, 2012; Hoffman & Ehrenfeld, 2015). Yet, research that spans levels of analysis is necessary to address sustainability issues both from an ecological standpoint (Rockström et al., 2009) and from a business perspective (Hahn et al., 2015; Starik & Rands, 1995; Whiteman et al., 2013). In this brief review we consider three broad analytical levels; the individual, organisational, and systemic, and these are reflected in the ordering of the chapters in this book. To end we briefly elaborate how and why we integrated a geographical perspective to support our analysis of the key issues raised in the book.

At the individual level, the relationship between sustainability and business can be played out by individuals within the boundaries of the business (e.g., owners, managers, employees, members, or volunteers) or those outside, such as consumers or citizens. The personal preferences and values of individuals can influence how decision-makers make sense and implement sustainability actions within a business (Bansal, 2003; Cordano & Frieze, 2000; Hahn et al., 2015). Thus, those internal to the organisation can significantly influence how a business responds. Research at this level of analysis has important implications for businesses, and development around this level of analysis has accelerated in recent years. Outside the organisation, individuals such as consumers and/or citizens have an important role to play. These individuals shape organisations as key stakeholders (Freeman, 1984) and can assert influence through preferences and social licence to operate.

The organisational, or business-level perspective, is central to and has dominated research in sustainability and business to date. This is perhaps unsurprising given that the field has largely developed from a concern about the relationship between business action and environmental and social issues (Bondy & Matten, 2012; Starik & Rands, 1995). At the organisational-level businesses use numerous natural processes in the creation, manufacture, and sales of products and services. Thus, businesses both affect and are affected by natural systems in a way that is inextricable (Whiteman et al., 2013). Research has examined this relationship from many perspectives as outlined in the previous section.

At the systemic level, the impact of business activity on planetary and social systems is an important focus. Thus, a system-level perspective moves beyond individuals and singular businesses and considers the contribution of business to a more sustainable society at large (Whiteman et al., 2013), and the contribution of business to a viable economy, sustainable society, and healthy ecosystems (Hahn et al., 2015). Conversely, research at this level also considers how the wider system affects the business. For example, Linnenluecke, Griffiths, and Winn (2012) make the argument that businesses are inherently vulnerable to the impacts of climate change and other extreme weather events. Thus, within the systems perspective it is necessary to consider the bi-directional impacts in the relationship between business and sustainability issues. This perspective echoes that of Starik and Rands (1995) who describe the relationships between levels as an open system, where each level affects and is affected by its relationship with other levels in the system; which in turn, leads to a systems perspective.

Finally, in this book we incorporate a diverse range of geographical perspectives to extend our understanding of the key issues, concepts, and themes pertaining to sustainability in business. Acknowledging the role of geography – the study of the interaction between people and the environment – in the field of sustainability and business helps us better understand the different dimensions of the human–environment relationship (e.g., cultural political, economic, urban) across scales, and between space and place in diverse contexts. This in turn initiates a deeper insight into why things happen where they do. To date, geographers have highlighted the intrinsic value of this approach in research investigating transitions towards more sustainable economies and societies (Bridge, Bouzarovski, Bradshaw, & Eyre, 2013; Hansen & Coenen, 2015). For the authors in this book, identifying and understanding impacts at, and across, different geographical scales are particularly important considerations when determining the impact of business activity on planetary and social systems – referred to as systemic-level impacts.

Further to this we recognise the importance and timeliness of the emerging decolonisation agenda within the scientific community. Decolonisation is the challenge to colonial (mainly European) structures, systems, and knowledge with the aim to work proactively to rebalance unequal power dynamics. Banerjee (2021) argues that colonial discourses inform our understanding of a range of topics associated within the field of sustainability of business, such as ecological sustainability, corporate social responsibility, stakeholder theory, and international business management. While the authors in this book appreciate the challenge of genuine and meaningful disentanglement from colonial discourses at this time, our modest contribution to the decolonisation agenda has been to seek out and integrate knowledge, insights, and case studies from outside of the Global North.

An outline of the chapters

The remainder of the chapters in the book map out some of the most pressing issues when examining the relationship between sustainability and businesses. Each chapter explores a specific issue that is crucial in gaining insight into the current issues facing businesses and identifies the current state of the art of research, identifies the most pressing research questions, and makes suggestions about possible methods. The chapters are broadly organised by the core focus of chapters and their dominant level of analysis. The chapters move from the most micro, or individual foci through to the most macro, or systemic foci. We have not, however, drawn distinct boundaries around the levels of analysis as many of the chapters consider issues and relationships that span levels and it would therefore be an artificial categorisation to draw distinct boundaries.

In Chapter 2, Onkila, Teerikangas, Koistinen and Mäkelä examine sustainability agency in business and present their findings of an interdisciplinary meta-review of relevant research. They argue that more research is needed at the micro-level to appreciate both individual- and collective-level action for sustainability and business. They call for interdisciplinary work of different types of actors in order to provide integration and engagement between different perspectives, theoretical lenses, and levels of analysis.

Chapter 3 from Lu, Chintakayala, Devinney, Young and Barkemeyer considers individuals from outside the organisation in their critical review of the socially responsible consumer. They explore research that examines why consumers often fail to translate pro-social intentions into pro-social consumption behaviours. They recommend future research that examines the complex systems and interconnections between consumers and other actors in order to more fully understand and explain socially responsible behaviour.

Our fourth chapter, from Unsworth and Pieniazek, moves towards the organisational-level of analysis and explores the concept of resilience in organisations as it relates to sustainability and business. They examine the similarities and differences between organisational resilience and organisational environmental sustainability. In doing so, the authors develop a multi-level framework and typology of organisational resilience in the face of sustainability issues, which serves as a basis for future research in this area.

Chapter 5 from Cutter, Trappmann and Krause reviews the current debate on work, employment, and climate change through the lens of 'just transition'. While focused on the labour-workforce, this chapter takes a systemic perspective and considers interrelationships across multiple levels of analysis. The authors argue that the engagement of workers is key in the development and implementation of strategies to reach net-zero carbon emission targets. They highlight some of the most pressing research questions and suggest case studies across companies, regions, and sectors will be necessary to inform policy-making to ensure that decarbonisation happens in a just way.

In Chapter 6 the authors Matoh, Roelich and Pinkse examine business models for sustainability, a topic gaining increasing attention and interest by organisational scholars. Their critical analysis shows that technological changes and developments are the main drivers of business model innovation for sustainability across three different industries. To better understand business model innovation for sustainability, the authors argue for multiple case studies and research across longitudinal time periods.

The extent and scale of sustainability impacts associated with business supply chains has garnered considerable academic and practitioner attention in recent years. It is estimated, for example, that eight supply chains alone – food, construction, fashion, fast-moving consumer goods (FMCG), electronics, automotive, professional services, and freight – account for half of all global annual greenhouse gas emissions. Reflecting on key debates in the Green Supply Chain Management (GSCM) literature, in Chapter 7 Wong and Zhu argue for a multi-tiered and multi-stakeholder perspective to drive forward a more progressive and challenging research agenda.

A key consideration for anyone undertaking sector- or industry-specific research – from a degree-level dissertation to doctoral or postdoctoral studies – is the research process to alight upon projects which are original and effective in enhancing the sustainability of their chosen sector. Drawing on their own experiences researching the UK and USA construction sectors, respectively, in Chapter 8 Owen and Francisco set out a practical step-by-step approach to identifying sector-specific research questions, followed by thoughts on how to answer those research questions.

Scholars argue that the transition to a low carbon future will require intensive mining for metals and minerals for green technologies. Many of the existing challenges associated with mining – social conflict, land dispossession, gender-based violence, and environmental destruction in the Global South – are exacerbated because of the increased demand for metals and minerals. In Chapter 9, Smith, Van Alstine and Ofori explore change and continuity in mining for the low carbon transition and discuss the challenges of conflict minerals and artisanal mining. They argue for a justice-focused research agenda for the extractive industries that draws across disciplines and is sensitive to the complexities of local contexts and power relations between actors.

In a useful complement to the previous chapter's focus on sustainability transitions in the extractive industries, in Chapter 10 Quartey describes a research agenda towards sustainability for small and medium-sized enterprises (SMEs) in Africa. SMEs play an instrumental role in the socio-economic development of many African countries yet there is considerable uncertainty as to how SMEs integrate sustainability principles within their business models. Quartey summarises SME characteristics, challenges, and contributions towards sustainable development and sets out a research agenda around the themes of institutions, technologies, business ethics, control systems/managerial behaviours, and indigenous management.

Solomon, Russell and Papargyropoulou present a review of sustainability management tools in Chapter 11. The outcome of their analysis suggests that the rising number of tools has led to increasing complexity and diversity in how businesses acknowledge and communicate their progress on sustainability issues. They argue that more needs to be done to consolidate this divergent field in order to facilitate consistency and comparability. Examining research on auditing and reporting assurance, the authors point to a developing field of research, where the findings have direct implications on organisational practice.

In an increasingly digital and interconnected world, businesses and organisations are adopting digital products and services to support their strategic and operational objectives. Despite the growth and continued innovation in these technologies, there is considerable uncertainty over how such innovations tackle sustainability issues at different scales. In Chapter 12, Padfield, Dales, Mishra and Smith provide a high-level overview of the sustainability impacts associated with digital innovation, and offer a way forward in terms of a research agenda.

In Chapter 13, Ghomi, Branicki, Brammer and Linnenluecke explore the ability of businesses to respond to rapid environmental changes like those brought about by the COVID-19 global pandemic. Using the concept of resilience, the authors use the pandemic to explore 'lessons learnt' and to examine how organisations can build resilience in times of crises. The authors offer up reflections and research questions to explore how organisations and society as a whole can build resilience in preparation for future rises and global challenges.

Drawing on emerging degrowth discourses, to complete the book in Chapter 14, Robra and Nesterova provide a distinctly critical perspective of the sustainability and business literature. Taking aim at arguments in favour of the 'business case for sustainability', the authors posit that businesses can never become fully sustainable as they are the fundamental form of capitalist economic organisation. From a degrowth perspective, 'true' sustainability is inherently incompatible with capitalism, meaning businesses are thus similarly incompatible with degrowth. Robra and Nesterova present the case for a research agenda on alternative forms of economic organisation, which leaves behind any attempt to find a common ground between degrowth and capitalism, its structures and agents.

References

Banerjee, S. B. (2012). Critical perspectives on business and the natural environment. In P. Bansal & A. J. Hoffman (Eds.), *The Oxford Handbook of Business and the Natural Environment* (pp. 572-590). Oxford, UK: Oxford University Press.

Banerjee, S. B. (2021). Decolonizing management theory: a critical perspective. *Journal of Management Studies*, 59(4), doi:10.1111/joms.12756.

Bansal, P. (2003). From issues to actions: the importance of individual concerns and organizational values in responding to natural environmental issues. *Organization Science*, 14(5), 510-527.

Bergman, M. M., Bergman, Z., & Berger, L. (2017). An empirical exploration, typology, and definition of corporate sustainability. *Sustainability (Switzerland)*, 9, 1-13.

Bergquist, A.-K. (2019). Business and sustainability. In T. da Silva Lopes, C. Lubinski & H. J. S. Tworek (Eds.), *The Routledge Companion to the Makers of Global Business* (pp. 546-563). London, UK: Routledge.

Bondy, K., & Matten, D. (2012). The relevance of the natural environment for corporate social responsibility research. In P. Bansal & A. J. Hoffman (Eds.), *The Oxford Handbook of Business and the Natural Environment* (pp. 519-536). Oxford, UK: Oxford University Press.

Bowen, H. R. (1953). *Social Responsibilities of the Businessman*. New York: Harper & Brothers.

Brenton, T. (1994). *The Greening of Machiavelli: The Evolution of International Environmental Politics*. London, UK: Earthscan Publications.

Bridge, G., Bouzarovski, S., Bradshaw, M., & Eyre, N. (2013). Geographies of energy transition: space, place and the low-carbon economy. *Energy Policy*, 53, 331-340.

Brundtland, G. H. (1987). *Report of the World Commission on Environment and Development: Our Common Future*. WCED.

Carroll, A. B. (1999). Corporate social responsibility: evolution of a definitional construct. *Business and Society*, 38(3), 268-295.

Carroll, A. B., Lipartito, K. J., Post, J. E., & Werhane, P. H. (2012). *Corporate Responsibility: The American Experience*. Cambridge, UK: Cambridge University Press.

Carson, R. (1963). *Silent Spring*. London: H. Hamilton.

Chang, R. D., Zuo, J., Zhao, Z. Y., Zillante, G., Gan, X. L., & Soebarto, V. (2017). Evolving theories of sustainability and firms: history, future directions and implications for renewable energy research. *Renewable and Sustainable Energy Reviews*, 72, 48-56.

Cordano, M., & Frieze, I. H. (2000). Pollution reduction preferences of U.S. environmental managers: applying Ajzen's theory of planned behavior. *Academy of Management Journal*, 43(4), 627-641.

Cormier, D., Magnan, M., & Van Velthoven, B. (2005). Environmental disclosure quality in large German companies: economic incentives, public pressures or institutional conditions? *European Accounting Review*, 14, 3-39.

Costanza, R., D'Arge, R., de Groot, R., Farber, S., Grasso, M., Hannon, B., ... van den Belt, M. (1997). The value of the world's ecosystem services and natural capital. https://royalroads.on.worldcat.org/oclc/4592801201. In *Nature*, 387, 253-260.

Crane, A., Matten, D., & Spence, L. J. (2008). *Corporate Social Responsibility: Readings and Cases in a Global Context*. London: Routledge.

Daly, H. E. (1998). The return of Lauderdale's Paradox. *Ecological Economics*, 25, 21-23.

Fombrun, C. (2005). Building corporate reputation through CSR initiatives: evolving standards. *Corporate Reputation Review*, 8, 7-11.

Forbes, L. C., & Jermier, J. M. (2012). The New corporate environmentalism and the symbolic management of organizational culture. In P. Bansal & A. J. Hoffman (Eds.), *The Oxford Handbook of Business and the Natural Environment* (pp. 556-571). Oxford, UK: Oxford University Press.

Freeman, R. E. (1984). *Strategic Management: A Stakeholder Approach*. Boston, MA: Pitman.

Hahn, T., Pinkse, J., Preuss, L., & Figge, F. (2015). Tensions in corporate sustainability: towards an integrative framework. *Journal of Business Ethics*, 127(2), 297-316. doi:10 .1007/s10551-014-2047-5.

Hansen, T., & Coenen, L. (2015). The geography of sustainability transitions: review, synthesis and reflections on an emergent research field. *Environmental Innovation and Societal Transitions*, 17, 92-109.

Hart, S. L. (1995). A natural-resource-based view of the firm. *Academy of Management Review*, 20(4), 986-1014.

Hoffman, A. J. (2001). *From Heresy to Dogma: An Institutional History of Corporate Environmentalism*. Palo Alto, CA: Stanford University Press.

Hoffman, A. J., & Bansal, P. (2012). Retrospective, perspective, and prospective: introduction to the *Oxford Handbook on Business and the Natural Environment*. In P. Bansal & A. J. Hoffman (Eds.), *The Oxford Handbook of Business and the Natural Environment* (pp. 3-25). Oxford, UK: Oxford University Press.

Hoffman, A. J., & Ehrenfeld, J. R. (2015). The fourth wave: management science and practice in the age of the Anthropocene. In S. A. Mohrman, J. O'Toole, & E. E. Lawler (Eds.), *Corporate Stewardship: Achieving Sustainable Effectiveness* (pp. 228-246). London, UK: Routledge.

IPCC. (2022). Climate change 2022: impacts, adaptation, and vulnerability. In D. C. R. H.-O. Pörtner, M. Tignor, E. S. Poloczanska, K. Mintenbeck, A. Alegría, M. Craig, S. Langsdorf, S. Löschke, V. Möller, A. Okem, & B. Rama (Eds.), *Contribution of Working Group II to the Sixth Assessment Report of the Intergovernmental Panel on Climate Change*. Cambridge: Cambridge University Press.

Jermier, J. M., Forbes, L. C., Benn, S., & Orsato, R. J. (2006). The new corporate environmentalism and green politics. In S. R. Clegg, C. Hardy, T. Lawrence, & W. R. Nord (Eds.), *The SAGE Handbook of Organization Studies* (pp. 618-650). London: Sage Publications.

Jones, G. (2017). *Profits and Sustainability: A Global History of Green Entrepreneurship*. Oxford, UK: Oxford University Press.

Kassinis, G. (2012). The value of managing stakeholders. In P. Bansal & A. J. Hoffman (Eds.), *The Oxford Handbook of Business and the Natural Environment* (pp. 83-100). Oxford, UK: Oxford University Press.

Kreps, J. T. (1940). *Measurement of the Social Performance of Business*. Washington DC: US Government Printing Office.

Kurucz, E., Colbert, B., & Wheeler, D. (2008). The business case for corporate social responsibility. In A. Crane, A. McWilliams, D. Matten, J. Moon, & D. Siegel (Eds.), *The Oxford Handbook of CSR* (pp. 83-112). Oxford, UK: Oxford University Press.

Linnenluecke, M. K., & Griffiths, A. (2013). Firms and sustainability: mapping the intellectual origins and structure of the corporate sustainability field. *Global Environmental Change*, 23, 382-391.

Linnenluecke, M. K., Griffiths, A., & Winn, M. (2012). Extreme weather events and the critical importance of anticipatory adaptation and organizational resilience in responding to impacts. *Business Strategy and the Environment*, 21(1), 17-32. doi:10 .1002/bse.708

Lozano, R., Carpenter, A., & Huisingh, D. (2015). A review of 'theories of the firm' and their contributions to corporate sustainability. *Journal of Cleaner Production*, 106, 430-442.

Margolis, J., Elfenbein, H., & Walsh, J. (2009). Does it pay to be good? A meta-analysis and redirection of research on the relationship between corporate social and financial performance. *SSRN*, https://ssrn.com/abstract=1866371 or http://dx.doi.org/10 .2139/ssrn.1866371.

McWilliams, A., & Siegel, D. (2001). Corporate social responsibility: a theory of the firm perspective. *Academy of Management Review*, 22, 823-886.

Moon, J., Crane, A., & Matten, D. (2005). Can corporations be citizens? Corporate citizenship as a metaphor for business participation in society. *Business Ethics Quarterly*, 15, 427-451.

Oldfield, F., Barnosky, A. D., Dearing, J., Fischer-Kowalski, M., McNeill, J., Steffen, W., & Zalasiewicz, J. (2014). The Anthropocene review: its significance, implications and the rationale for a new transdisciplinary journal. *The Anthropocene Review*, 1(1), 3-7.

Rockström, J., Steffen, W., Noone, K., Persson, Å., Chapin III, F. S., Lambin, E., ... Schellnhuber, H. J. (2009). Planetary boundaries: exploring the safe operating space for humanity. *Ecology and Society*, 14(2), art 32.

Scherer, A. G., Palazzo, G., & Matten, D. (2009). The changing role of business in a global society: new challenges and responsibilities. *Business Ethics Quarterly*, 19, 327-347.

Shaw, D., & Clarke, I. (1999). Belief formation in ethical consumer groups: an exploratory study. *Marketing Intelligence & Planning*, 17, 109-119.

Starik, M., & Rands, G. P. (1995). Weaving an integrated web: multilevel and multisystem perspectives of ecologically sustainable organizations. *Academy of Management Review*, 20(4), 908-935.

Whiteman, G., Walker, B., & Perego, P. (2013). Planetary boundaries: ecological foundations for corporate sustainability. *Journal of Management Studies*, 50(2), 307-336. doi:10.1111/j.1467-6486.2012.01073.x.

Windsor, D. (2006). Corporate social responsibility: three key approaches. *Journal of Management Studies*, 43, 93-114.

Wu, M.-L. (2006). Corporate social performance, corporate financial performance, and firm size: a meta-analysis. *The Journal of American Academy of Business*, 8, 163-171.

2 Sustainability agency in business: an interdisciplinary review and research agenda

Tiina Onkila, Satu Teerikangas, Katariina Koistinen and Marileena Mäkelä

1.   Introduction

Multiple actors, such as managers, activists, employees and customers, in and around business have identified significant challenges related to sustainability and have started to argue for the need for change, if not paradigm shifts, in the ways of doing business. Taking a closer look, the role of human action is crucial when pursuing sustainability in business. Across the social sciences, the role of human actors is often approached via the concept of agency, which refers to the human capacity to act and to make a difference (Giddens, 1984; Dietz and Burns, 1992; Bandura, 2006). Further, agency is considered as being intentional and reflexive (Dietz and Burns, 1992; Bandura, 2001). Understanding the relation between business and sustainability requires knowledge of the role of agency in shaping this relation towards greater degrees of un/sustainability. In recent years, the need to pay attention to human action as an enabler of change has been noted (Aguinis and Glavas, 2012). This involves asking questions, such as what are or can be the contributions of actors to sustainability, as well as why and how they contribute to sustainability in business (Pesch, 2015; Koistinen, 2019).

While studies have started to highlight the role and importance of human agency towards sustainability change in business, upon closer examination, no single discipline owns this field of research. In other words, this appreciation is fragmented across disciplines, theories, literatures, phenomena and journals. This mirrors the situation regarding the study of sustainability agency in general, beyond business contexts. Indeed, based on our meta-review of research on different forms and types of sustainability agency across disciplines, levels of analysis, sectors and contexts, we observed the difficulty in finding a discipline with a clear focus on or ownership of the study of sustainability actors (Teerikangas et al., 2021a). While the above review provided an interdisciplinary perspective on sustainability agency, there is, further, a need for a more in-depth and systematic understanding of sustainability agency in business contexts and the role of actors in making business more sustainable. Additionally, we perceive the need for more knowledge of how sustainability actors act, how contextual factors shape their approaches and how they aim to influence business towards sustainability.

In this chapter, we take up this challenge. Our aim is to provide an integrative view on sustainability agency in business, particularly regarding the questions of who the relevant actors are and how they act. To appreciate sustainability agency in business settings and for the purposes of this chapter, three disciplines are of interest: sustainability science, management studies and corporate social responsibility (CSR). For one, sustainability science appreciates agency in the broader, systemic context of sustainability transitions (Fischer and Newig, 2016; Koistinen et al., 2019; Koistinen and Teerikangas, 2021). For another, two disciplines can be pointed out as having focused on sustainability agency as it relates to business: management studies and CSR research. In this chapter, we thus review extant research on sustainability agency, as it relates to business, across the sustainability transitions, management studies and CSR streams of literature. In so doing, our main contribution is in offering a meta-review of prior research on sustainability agency in business settings, a typology of sustainability actors, as well as pointers towards future research.

We have organised this chapter as follows: in Section 2, we introduce the concept of sustainability agency. In Section 3, we introduce our research method and review how it has been studied in the sustainability transitions, management studies and CSR streams of literature, followed by a synthesis of this understanding. Finally, in Section 4, we discuss future research directions arising from this analysis.

2.   What is sustainability agency?

Agency is a classic concept across the social sciences, particularly used in, but not limited to sociology (Emirbayer and Mische, 1998). For the purposes of this chapter, we proceed to a selected overview of classics of agency, moving thereafter to appreciating and then defining sustainability agency.

Agency can be defined as an individual's or a collective's capacity to act, though the bulk of research tends to adopt the individual-level focus (Giddens, 1984;

Dietz and Burns, 1992). Over time, the concept of agency has been viewed and defined across the social sciences in different ways, using a variety of theoretical perspectives (Bandura, 2002; Emirbayer and Mische, 1998; Koistinen, 2019; Teerikangas et al., 2021a). For example, according to Bandura (2002), agency refers to intentionally influencing one's functioning and life circumstances. Based on Stones (2005), Sherwin (2009) and Tourish (2014), Koistinen (2019) summarises agency as intentional action, involving the possession of power. Thus, it can be defined as the 'ability to engage in purposeful action' (Tourish, 2014, p. 87) and 'having the capacity to take an action' (Tourish, 2014, p. 80). Although multiple non-human forms of agency are recognised (Latour, 2005; Jokinen et al., 2021), in this chapter, we focus on human agency. Such a definition as the capacity to act (Bandura, 2001) reflects free will and determinism (Emirbayer and Mische, 1998), thus characterising the historical evolution of humankind.

Similarly, the concepts of agency or agent can be found in both the CSR and the sustainability transition streams of literature. In these literatures, the focus is on individual and/or collective agents and their agency in furthering CSR and sustainability transitions. Taking a closer look, within the framework of corporate sustainability, the term agent refers to individual and collective actors as participants in purposive actions in their attempt to either prevent or generate change (Bos et al., 2013; Fischer and Newig, 2016). In the field of sustainability transitions, agency can be defined as an actor's behaviour with regard to such change (Loorbach, 2007). Koistinen (2019) stresses that agency furthering sustainability transitions may also be a collective action and a social phenomenon that is shaped by sociocultural contexts (Billett, 2006; Eteläpelto et al., 2013). Indeed, individual and social agency are mutual and should be considered as intertwined (Billett, 2006).

Concerning agency in sustainability transitions, Pesch (2015) emphasises that the questions of why individual agents make certain decisions, why they have certain motivations and how these motivations can be influenced remain largely unanswered. He also views individual agents as parts of larger societal and institutional realms, motivated by different contextual factors. He calls for a more in-depth understanding of what drives people at different societal levels to embed into sustainability change.

Based on an interdisciplinary literature review, sustainability agency is proposed as an umbrella concept to incorporate the diversity of actors engaging in sustainability work (Teerikangas et al., 2021a). The concept of sustainability agency offers an integrative take, across disciplines, on the phenomenon of active sustainability agency. In so doing, the concept offers an umbrella term for sustainability actors operating at different levels of analysis, contexts, while studied with varying theoretical lenses across disciplines, thereby encapsulating different actor types. The concept therefore encapsulates individual, activist and relational forms of agency, as well as governance as a mode of agency. Further, the concept of agency is often betted in opposition to and in tandem with surrounding social and societal structures, and a classic question in sociology has been which one matters, agency or structure (Ritzer, 2005). Indeed, while the literatures abound with similar concepts, for example sustainability performance and sustainability behaviours, the uniqueness of sustainability agency is its operating at an aggregate level, encapsulating various forms of agency. It has focus on intentionality and active change orientation, both at individual and collective levels, towards sustainable futures. While for example the concept of sustainability behaviour shares the interdisciplinary nature of sustainability agency, it lacks orientation towards intentionality and active change creation – instead, sustainability behaviour can also maintain unsustainable practice and rather mechanically repeat prior traditions with no focus on change, and be open to the influence of others (Ketron and Naletelich, 2019). As another example, while sustainability agency pays attention at the full processes, for example sustainability performance is oriented on results of such action and characterised by collective-level focus (see e.g. Papoutsi and Sodhi, 2020).

3.   Methods

We focused on reviewing three streams of literature in order to gain an appreciation of sustainability agency in business: (1) sustainability transition literature discusses the role of agency in sustainability transition in general; and (2) management studies given its generic focus on management phenomena; and (3) CSR literature given its explicit focus on responsible business. As each stream of literature differs in its focus on agency towards sustainable business, we conducted three parallel reviews, using selected methodologies and search words for each, in order to obtain a thorough coverage and an understanding of the status of sustainability agency research across the studied streams of literature.

1. In sustainability transition literature, literature search was done in November 2018 in the Scopus database, using the selected terms "*agency*" AND *sustainability transitions*' and "*agent*" AND *sustainability transitions*' in paper abstracts, titles or keywords. We limited the search to cover the period of 2014–2018. The search resulted in 270 publications. After exclud-

ing papers that did not focus on agency and sustainability transitions, 77 journal articles were included in the final sample for review.

2. We continued our search of sustainability agency in mainstream management journals. Our focus was on the leading 16 journals in the field (i.e. *Journal of Management Studies, Journal of Management, British Journal of Management, Academy of Management Journal, Academy of Management Review, Academy of Management Annals, Organization Studies, Research Policy, Human Relations, Administrative Science Quarterly, International Journal of Management Reviews, Journal of International Business Studies, Long Range Planning, Organization Science, Strategic Management Journal,* and *Strategic Organization*). As the search with the exact term 'sustainability agency' led to few meaningful results, the search was enlarged to encompass terms related to sustainability, including sustainability, CSR, responsibility, climate change, and energy as well as proxies for agency including agent/agency, manager/professional/employee, grassroots, community, niche, activism, social movement, non-governmental organisations (NGO), and social entrepreneur. After cross-checking the full sample, the final sample reviewed included 150 papers published in the period of 1992-2020.

3. Finally, we reviewed leading CSR journals in order to appreciate the state of the art of research on sustainability agency. The reviewed journals included *Organization and Environment, Business Strategy and the Environment, Journal of Business Ethics* and *Business and Society*. Our search was based on the words agency, championship, activism, advocacy and pioneer. It was conducted in November 2019. This resulted in 281 articles, in which the search terms were mentioned in the paper title, abstract or keywords. After exclusion based on the centrality of the search terms in the articles and the meanings in which the search terms were used, the results were narrowed down to 88 articles.

We applied different literature search methodologies for each of the three streams of literature. This choice reflected the fact that, upon conducting the searches, we observed each stream to address and discuss sustainability agency in different ways. Put differently, each stream of research appeared to bear a different degree of maturity in the study of sustainability agency. While the sustainability transition literature has an explicit, ongoing discussion using the term 'agency', such a discussion is missing from the management studies and CSR literatures. Taking a closer look, in the latter streams, the search with the term 'sustainability agency' or 'sustainability agent' led to few, if any, results. Therefore, as researchers, we needed to revert to different search methodologies and search words for each of the three streams of literature. In CSR literature, where the relevant discussion was more extensive, the search could

Table 2.1   Methodologies for literature reviews

Stream of literature	Search words	Target journals	Number of articles
Sustainability transition	agency AND sustainability transitions and agent AND sustainability transitions	Searches were targeted at sustainability science literature and especially the discipline of sustainability transitions, with no specific journal limitation.	77 articles
Management	terms related to sustainability, including sustainability, CSR, responsibility, climate change, and energy as well as proxies of agency including agent/ agency, manager/ professional/ employee, grassroots, community, niche, activism, social movement, NGO, and social entrepreneur	16 leading journals in the management (*Journal of Management Studies, Journal of Management, British Journal of Management, Academy of Management Journal, Academy of Management Review, Academy of Management Annals, Organization Studies, Research Policy, Human Relations, Administrative Science Quarterly, International Journal of Management Reviews, Journal of International Business Studies, Long Range Planning, Organization Science, Strategic Management Journal,* and *Strategic Organization*).	150 articles
CSR	agency, championship, activism, advocacy and pioneer	5 leading CSR journals (*Organization and Environment, Business Strategy and the Environment, Journal of Business Ethics, Business and Society* and *Journal of Cleaner Production*).	88 articles

be done based on terms similar to agency, but in the management journals, also actor groups had to be included in search words. Further, the timescale of our reviews varied depending the existence of prior literature reviews. To this end, the sustainability transition review was done in 2019, covering only the years 2014–2018, given that a prior review exists that covers years leading to 2014 (Fischer and Newig, 2016). In management and CSR literatures, no prior reviews were found, and hence a longer timespan was adopted. The different methods for our literature reviews are summarised in Table 2.1.

In the following subsections, we proceed to detailing the results of our review per stream of research before integrating the findings towards an appreciation of sustainability agency in business contexts.

4.   Findings: reviews of sustainability agency research in business

4.1   Review of agency in the literature on sustainability transitions

We began our search in the field of sustainability science, where the bulk of theorising on various forms of sustainability transitions occurs (Fischer and Newig, 2016; Köhler et al., 2019). We reviewed extant literature on agency in sustainability transitions between 2014 and 2018, as we were building on the work of Fischer and Newig (2016), who reviewed the role of actors in sustainability transitions between 1995 and 2014. For another, as the world has experienced the outbreak of various sustainability actions led by individuals and collectives, and the study of agency in sustainability transitions has burgeoned since 2014, we sought to appreciate recent developments in the field. For a thorough overview of the findings of our review 2014-2018, please see Koistinen and Teerikangas (2021).

We noticed an increasing amount of research on agency in the sustainability transition literature in each year from seven publications in 2014 to 25 publications in 2018. Our observation was that the terminology was scattered as regards the terms agent, actor and agency (e.g. Antadze and McGowan, 2017). A closer look revealed that the contemporary transition literature on agency emphasises three themes: (1) governance (e.g. Klinke, 2017), (2) agent typologies (e.g. Avelino and Wittmayer, 2016), and (3) calls for richer views regarding agency (e.g. van der Vleuten, 2018). First, we observed that the majority of transition studies vis-à-vis agency discussed governance, politics, power of agency or institutions and agency. Second, our findings showed that the transition literature typically conceptualised agents via various typologies. These typologies include categories such as change agency, niche formation, incumbents and strategic agency. Third, we noted that emerging topics in the literature related to persistent calls for richer perspectives regarding agency involved in sustainability transitions, whether in the form of behavioural sciences or views adopted from socio-ecological research on sustainability.

Despite the increasing interest in agency in sustainability transitions, these studies are scattered and set amid various theoretical underpinnings, such as institutional theory, structuration theory or practice theory (e.g. Kuhmonen, 2017; Stephenson, 2018; Koistinen and Teerikangas, 2021). Upon closer examination, the literature tends to be set on a persistent debate emphasising either the system (e.g. de Gooyert et al., 2016) or the agent (e.g. Bögel and Upham, 2018) as bearing primary importance in the making of sustainability transitions. This mirrors the classic agency-structure question in sociology (e.g. Giddens, 1984; Emirbayer and Mische, 1998).

4.2   Review of agency and sustainability in the literature on management

Our review of the management studies literature led us to identify numerous types of actors, that is, forms of agency geared towards sustainable business. To begin with, there are actors in the broader societal and institutional environment affecting a firm's sustainability strategy including transnational players, national governments and their environmental policy and regulation efforts, government-affiliated intermediary organisations, regional players within countries, various stakeholders, investors, financial advisors, investment funds, and a firm's owners (e.g. Patriotta et al., 2011; Crouch, 2006; Doh et al., 2010). Second, there is increasing interest in the study of incumbent firms as sustainability actors. In this realm, the factors influencing a firm's sustainability strategy are under study, alongside their performance effects. The role of political CSR is studied (Scherer and Palazzo, 2011), and a number of authors posit firms as operating as both incumbents and activists towards sustainability (e.g. Berggren et al., 2015). Beyond regular performance metrics, questions regarding responsible innovation, new market creation, business models, and growth via mergers and acquisitions (M&As) are raised (Peloza, 2009; Pitelis, 2009). Projects are identified as potential vehicles towards sustainability. The means of managing a firm from a sustainability perspective are questioned, for example, as regards the organisation's logic, its decision-making model, procurement contracts, supply chain management, and the way in which it orchestrates its CSR strategy. In the study of organisations, questions of organisational culture and identity are studied.

Third, questions of collaborative agency have been raised. In this respect, beyond the firm itself, questions of cross-sector partnerships and clusters are studied. Also, questions of shared or collective governance are of interest. Fourth, sustainability agency has been identified to occur within incumbent organisations. In this regard, the focus has been on specific individual roles in the organisational hierarchy, such as board members, chief executive officers, executives, decision-makers, managers, sustainability or CSR professionals and managers, as well as employees (Whiteman and Cooper, 2000; Shepherd et al., 2013; Kim et al., 2017; Mitra and Buzzanelli, 2017). The notion of active agency is studied via the notions of embedded agency (Fan and Zietsma, 2017) and institutional entrepreneurship. In parallel, there is some, though scant interest towards consumers, be it as regards citizen users or consumer behaviour.

Fifth, a number of active organisational sustainability actors are studied by management scholars. Such actors include sustainable entrepreneurs, social entrepreneurs, communities, non-governmental organisations, social movements, grassroots organisations, activists (whether climate change, civic, institutional or employee activists), communities of place, communities of action, community-based enterprises, neighbourhoods, and base of the pyramid actors (e.g. Khan et al., 2010; Markman et al., 2016).

In summary, while the term agency itself was not actively in use, numerous actor types, be it individuals, organisations or collectives, could be identified in this literature. Further, the interest towards their study is increasing in recent years. Upon closer look, the study of each actor type is set amid a specific phenomenon-based literature and theoretical debate, with little cross-fertilisation across these literatures. As such, our review offers an emerging integrative perspective on the plethora of sustainability actor types studied in this literature.

4.3   Review of agency in the literature on CSR

The reviewed research connected multiple actors with sustainability. These actors were both individuals and organisations, including companies (multinational, small-/medium-sized and family-owned companies), individuals belonging to organisations and societies, managers and leaders, consumers, pioneers and champions in different contexts, public sector actors, shareholders, communities and social movements, NGOs and environmental activists, labour unions and stakeholders as a whole.

The analysis focused on the question of 'how sustainability actors act towards sustainability'. In the studies' main findings, we noticed that a majority of the studies focused on how the actors aimed to influence others to promote sustainability. Studies on how the actors themselves acted were less prevalent. Based on the dominant trend, we focused on identifying different influence strategies – how sustainability actors aimed to influence others and thereby to promote sustainability. We identified two main strategies that dominated these studies: influential and co-productive. Regarding the influential strategies, the actors aimed to use their power to promote sustainability, while as regards co-productive strategies, they aimed at collective action to promote sustainability.

To begin with influential strategies, these studies focused on how the actors aimed to influence others by promoting or demanding sustainability. This entailed both direct and indirect strategies. On the one hand, direct strategies were the focus of those studies that examined the direct use of power by a certain actor or actor group (Juravle and Lewis, 2009; Georgallis, 2017; Walls and Berrone, 2017). These studies dealt with how to convince others to pursue sustainability; thus, the concept of power did not refer only to visible sources of power but also to many invisible sources of power and ways to use it (O'Rourke, 2003; Galbreath, 2010). For example, Georgallis (2017) showed that social movements are able to influence the expectations that key stakeholders have about firms' social responsibility, making corporate social initiatives more attractive. On the other hand, indirect strategies were addressed by those studies that investigated the indirect use of power by a certain actor or actor group. These studies dealt with the power of networking, the power of language use and the power of interaction, thus entailing invisible ways of exercising power (Lewis and Juravle, 2010; Hancock and Nuttman, 2014; Lorek and Spangenberg, 2014; Sarasini and Jacob, 2014; Peattie and Samuel, 2018). For example, Sarasini and Jacob (2014) showed how managers may reconstruct and sometimes refute the pressures for climate action, and thus shape how it is approached among other actors.

Moving onto co-productive strategies, they dealt with the collective action to promote sustainability and described how one actor group acted in collaboration with others towards sustainability change. However, this type of studies was notably smaller in number than those focusing on influential strategies. In this type of studies, we identified two subthemes: studies highlighting the role of collaboration in the adoption of sustainable solutions (McLaughlin, 2012; Gauthier and Gilomen, 2016) and studies highlighting the limitedness of the single-actor approach (Green et al., 2000; Berry, 2003; Sonpar et al., 2009). For example, Gauthier and Gilomen (2016) showed how collective agency facilitates development towards sustainability and adoption of sustainable solutions. In summary, while the term agency itself was limitedly in use, numerous actor and actions types could be identified in this literature.

4.4   On the nature of sustainability agency in business

Our review of sustainability agency across the three disciplines highlights this topic as of increasing interest. In an effort to summarise and integrate the findings, we observe the following. To begin with, we find that the three disciplines have adopted different foci in their study of sustainability agency. First, sustainability science, particularly the sustainability transition literature, has addressed the role of agency and has increasingly started to pay both thematic and conceptual attention to the role of agency in the broader, systemic context of sustainability transitions (Fischer and Newig, 2016; Koistinen et al., 2019). Second, management research exhibits a wide array of different types of actors and how they may contribute to sustainability change (Teerikangas et al., 2018), while CSR research has shown how these different actors may use their influence to convince others on the cause of sustainability (Onkila et al., 2019).

All the while, the reviewed disciplinary fields seem to operate with relative independence from one another, leading to siloed approaches to the study of sustainability agency as it relates to business. Under such circumstances, obtaining an overview of 'what sustainability agency in business is' becomes an arduous and laborious task. It can even be argued that increasing within-disciplinary research may be counterproductive in the long term, when compared against efforts to synthesise this knowledge.

More alarmingly, we find further siloing within the disciplines. Indeed, our analysis shows how the research in each of the three fields has identified numerous actor types, each often discussed in a separate stream of literature. In other words, the research on sustainability agency in business is not only scattered across disciplines, but further, within disciplines across the study of numerous actor types and levels of analysis, each representing a literature area of its own. Indeed, there is little attempt at integration and synthesis within or across literatures, theoretical debates or disciplines. Subsequently, it is difficult to gain an overview of what sustainability agency in business is and what forms it takes. In this chapter, our contribution lies in bringing forth an integrative perspective to sustainability agency in business, as well as a typology of the relevant actors.

In seeking answers to the question of 'what sustainability agency is', researchers need to recognise that instead of the term 'agency', other terms reflecting active action towards the sustainability agenda are helpful proxies in identifying this literature. Such terms include, but are not limited to, the terms social entrepreneur, transnational standard setter, social movement and activism. Thus, numerous organisational and individual actors have been studied as active sustainability actors. Beyond firms, other organisations operating in the institutional environment, as well as various types of activist organisations, have been examined. Taking a closer look at individuals, the focus appears to be on the study of sustainability-active individuals working for incumbent firms, whether in managerial, professional or employee roles, who actively drive sustainability strategies. Studies on activist organisations focus on the organisation, instead of the individual(s) driving the organisation. Further, the role of collaboration in pursuing sustainable futures is advanced. In synthesis, our three-disciplinary review of the studies on active sustainability actors leads us to view sustainability agency as occurring either at individual or organisational levels by/in incumbent or activist organisations when they proactively and collaboratively pursue sustainable futures.

5.   Future research on sustainability agency in business

As discussed in this chapter, three disciplines have addressed sustainability agency in business settings: sustainability transition, management studies and CSR research. Going forward, we suggest that future research on sustainability agency in business is a prerequisite for an in-depth understanding of the role of agency in order to steer towards sustainable business.

Based on the three reviews, it appears that prior research has recognised a wide variety of sustainability actors. All the while, these pockets of research appear to operate in relative isolation and in siloes within and across disciplines. In the lack of integrative overviews (on these actors and their agentic practices) researchers and practitioners retain siloed and actor-based views of sustainability agency. While such an actor-focused perspective enables an in-depth appreciation of an individual actor type, it disallows comparisons, cross-fertilisation and integration of knowledge. Such state of affairs can be considered dire and discouraging at this time in the 2020s when the general public's sustainability awareness is increasing and the role of actors is increasingly called for. In the absence of integrative perspectives, the academic community can only provide partial answers. Going forward, we call for more interdisciplinary research on sustainability agency, with an aim to provide integration and engagement between different perspectives, theoretical lenses and levels of analysis. We have begun this work in drawing together an interdisciplinary perspective to the matter (Teerikangas et al., 2021a), but there is room and need for much further work. We argue that going forward, sustainability agency scholars need to work across disciplines, literature streams and theoretical bases. Cross-fertilisation and integration of knowledge areas is not only a means of contributing scientifically (Ladik and Stewart, 2008), but further, a societal service at a time when individual and organisational actors increasingly ponder about their role in ensuring sustainable futures. It is somewhat paradoxical that while much scientific research exists, as long as it is scattered, it is difficult for the societal audiences to reach this knowledge, and hence, to develop their sustainability agency, and appreciate how one actor type's agency relates to others, and how, collectively and collaboratively, sustainable futures can be enabled.

Prior literature has also offered analyses on how these actors act. However, the understanding seems dominated by the analysis of influence. We still know little about such questions as what is the variety of agency practices – what else does sustainability agency influence? Change agency is an integral part of sustainability agency (Teerikangas et al., 2021b), as incremental, radical and emerging changes are direly needed to integrate sustainability into business. How do these sustainability actors enforce sustainability, and what obstacles do they face? Additionally, responses to active agency are still poorly understood. How does agency trigger positive responses, and how may it encourage others to act? How can collective agency be enforced by individual agency, and what is the role of collaboration in sustainability agency? We suggest that future research should focus on studying the variety of agency practices in different contexts to gain a better understanding of the questions of how sustainability agents act. This would especially require applications of qualitative case studies in organisational settings and in stakeholder relations to enable the use of multiple data sources in order to understand the variety of agency practices and processes.

Furthermore, based on our reviews, we call for a micro-level understanding about the questions of why and when sustainability agents act. The literature has not yet offered an in-depth understanding of actors' experience in sustainability agency – what enforces agency, and what motivates action? We thus know little about the dynamics of individual-level and collective-level agency, including behaviours, motivations, actor attributes and contextual processes shaping agency. Specifically, understanding about bottom-up actions, such as changes initiated by employees within organisations, is still missing. More in-depth knowledge regarding these questions requires the inclusion of psychological theory and behavioural sciences, addressing in particular individuals' behaviours whether alone or as part of organisational and societal contexts. To further the understanding about how individual and collective agency develop in specific contexts, narrative and longitudinal methodologies are called for in order to find out what have been the motives for agency development and what contextual processes have shaped agency.

In turn, actors' motivations also include change-resisting practices. To understand the complex dynamics of creating business sustainability, passive and resisting actors should not be forgotten. Thus, future research could delve into the issue of actor rationales in the form of more passive and even resisting actors in business environments. Furthermore, the concept of power in relation to sustainability agency deserves more attention. What role does power play in building and overcoming resistance?

References

Aguinis, H. and A. Glavas (2012), 'What we know and don't know about corporate social responsibility: A review and research agenda', *Journal of Management*, **38** (4), 932-968.

Antadze, N. and K. McGowan (2017), 'Moral entrepreneurship: Thinking and acting at the landscape level to foster sustainability transitions', *Environmental Innovation and Societal Transitions*, **25**, 1-13.

Avelino, F. and J. Wittmayer (2016), 'Shifting power relations in sustainability transitions: A multi-actor perspective', *Journal of Environmental Policy Planning*, **18**, 628-649.

Bandura, A. (2001), 'Social cognitive theory: An agentic perspective', *Annual Review of Psychology*, **52** (1), 1-26.

Bandura, A. (2002), 'Selective moral disengagement in the exercise of moral agency', *Journal of Moral Education*, **31** (2), 101-119.

Bandura, A. (2006), 'Toward a psychology of human agency', *Perspectives on Psychological Science*, **1** (2), 164-180.

Berggren, C., T. Magnusson and D. Sushandoyo (2015), 'Transition pathways revisited: Established firms as multi-level actors in the heavy vehicle industry', *Research Policy*, **44** (5), 1017-1028.

Berry, G.R. (2003), 'Organizing against multinational corporate power in cancer alley: The activist community as primary stakeholder', *Organization & Environment*, **16** (1), 3-33.

Billett, S. (2006), 'Relational interdependence between social and individual agency in work and working life', *Mind, Culture, and Activity*, **13** (1), 53-69.

Bögel, P.M. and P. Upham (2018), 'The role of psychology in sociotechnical transitions literature: A review and discussion in relation to consumption and technology acceptance', *Environmental Innovation and Societal Transitions*, **28**, 122-136.

Bos, J., R. Brown and M.A. Farrelly (2013), 'A design framework for creating social learning situations', *Global Environmental Change*, **23** (2), 398-412.

Crouch, C. (2006), 'Modelling the firm in its market and organizational environment: Methodologies for studying corporate social responsibility', *Organization Studies*, **27** (10), 1533-1551.

de Gooyert, V., E. Rouwette, H. van Kranenburg, E. Freeman and H. van Breen (2016), 'Sustainability transition dynamics: Towards overcoming policy resistance', *Technological Forecasting and Societal Change*, **111**, 135-145.

Dietz, T. and T.R. Burns (1992), 'Human agency and the evolutionary dynamics of cult ure', *Acta Sociologica*, **35** (3), 187-200.

Doh, J.P., S.D. Howton, S.W. Howton and D.S. Siegel (2010), 'Does the market respond to an endorsement of social responsibility? The role of institutions, information, and legitimacy', *Journal of Management*, **36** (6), 1461-1485.

Emirbayer, M. and A. Mische (1998), 'What is agency?', *American Journal of Sociology*, **103** (4), 962-1023.

Eteläpelto, A., K. Vähäsantanen, P. Hökkä and S. Paloniemi (2013), 'What is agency? Conceptualizing professional agency at work', *Educational Research Review*, **10**, 45-65.

Fan, G.H. and C. Zietsma (2017), 'Constructing a shared governance logic: The role of emotions in enabling dually embedded agency', *Academy of Management Journal*, **60** (6), 2321-2351.

Fischer, L.B. and J. Newig (2016), 'Importance of actors and agency in sustainability transitions: A systematic exploration of the literature', *Sustainability*, **8** (5), 476.

Galbreath, J. (2010), 'Corporate governance practices that address climate change: An exploratory study', *Business Strategy and the Environment*, **19** (5), 335-350.

Gauthier, C. and B. Gilomen (2016), 'Business models for sustainability: Energy efficiency in urban districts', *Organization & Environment*, **29** (1), 124-144.

Georgallis, P. (2017), 'The link between social movements and corporate social initiatives: Toward a multi-level theory', *Journal of Business Ethics*, **142** (4), 735-751.

Giddens, A. (1984), *The Constitution of Society*, Cambridge: Polity.

Green, K., B. Morton and S. New (2000), 'Greening organizations: Purchasing, consumption, and innovation', *Organization & Environment*, **13** (2), 206-225.

Hancock, L. and S. Nuttman (2014), 'Engaging higher education institutions in the challenge of sustainability: Sustainable transport as a catalyst for action', *Journal of Cleaner Production*, **62**, 62-71.

Jokinen, A., J. Uusikartano, P. Jokinen and M. Kokko (2021), 'The interagency cycle in sustainability transitions', in Teerikangas, S., T. Onkila, K. Koistinen and M. Mäkelä (eds.), *Research Handbook on Sustainability Agency*, Cheltenham, UK and Northampton, MA, USA: Edward Elgar Publishing.

Juravle, C. and A. Lewis (2009), 'The role of championship in the mainstreaming of Sustainable Investment (SI): What can we learn from SI pioneers in the United Kingdom?', *Organization & Environment*, **22** (1), 75-98.

Ketron, S. and K. Naletelich (2019), 'Victim or beggar? Anthropomorphic messengers and the savior effect in consumer sustainability behavior', *Journal of Business Research*, **96**, 73-84.

Khan, R.F., R. Westwood and D.M. Boje (2010), '"I feel like a foreign agent": NGOs and corporate social responsibility interventions into Third World child labor', *Human Relations*, **63** (9), 1417-1438.

Kim, A., Y. Kim, K. Han, S.E. Jackson and R.E. Ployhart (2017), 'Multilevel influences on voluntary workplace green behavior: Individual differences, leader behavior, and coworker advocacy', *Journal of Management*, **43** (5), 1335-1358.

Klinke, A. (2017), 'Dynamic multilevel governance for sustainable transformation as postnational configuration', *Innovation: The European Journal of Social Science Research*, **30** (3), 323-349.

Köhler, J., F.W. Geels, F. Kern, J. Markard, A. Wieczorek, F. Alkemade, F. Avelino, A. Bergek, F. Boons, L. Fünfschilling, D. Hessk, G. Georg Holtz, S. Hyysalo, K. Jenkins, P. Kivimaa, M. Martiskainen, A. McMeekin, M.S. Mühlemeier, B. Nykvist, E. Onsongo, B. Pel, R. Raven, H. Rohracher, B. Sandén, J. Schot, B. Sovacool, B. Turnheim, D. Welch and P. Wells (2019), 'An agenda for sustainability transitions research: State of the art and future directions', *Environmental Innovation and Societal Transitions*, **31**, 1-32.

Koistinen, K. (2019), Actors in sustainability transitions. Doctoral dissertation. LUT University Press.

Koistinen, K. and S. Teerikangas (2021), 'The debate if agents matter vs. the system matters in sustainability transitions: A review of the literature', *Sustainability*, **13** (5), 2821.

Koistinen, K., S. Teerikangas, T. Onkila and M. Mäkelä (2019), 'The debate regarding agents and sustainability transitions – review of the literature', Paper presented at Corporate Responsibility Research Conference, 12-13 September, 2019, Tampere, Finland.

Kuhmonen, T. (2017), 'Exposing the attractors of evolving complex adaptive systems by utilising futures images: Milestones of the food sustainability journey', *Technological Forecasting and Societal Change*, **114**, 214–225.

Ladik, D.M. and D.W. Stewart (2008), 'The contribution continuum', *Journal of the Academy of Marketing Science*, **36** (2), 157–165.

Latour, B. (2005), *An Introduction to Actor-Network Theory. Reassembling the Social*, New York: Oxford University Press.

Lewis, A. and C. Juravle (2010), 'Morals, markets and sustainable investments: A qualitative study of "champions"', *Journal of Business Ethics*, **93** (3), 483–494.

Loorbach, D. (2007), *Transition Management: New Mode of Governance for Sustainable Development*, Utrecht, the Netherlands: International Books.

Lorek, S. and J.H. Spangenberg (2014), 'Sustainable consumption within a sustainable economy: Beyond green growth and green economies', *Journal of Cleaner Production*, **63**, 33–44.

Markman, G.D., M. Russo, G.T. Lumpkin, P.D.D. Jennings and J. Mair (2016), 'Entrepreneurship as a platform for pursuing multiple goals', A Special Issue on Sustainability, Ethics, and Entrepreneurship. *Journal of Management Studies*, **53** (5), 673–694.

McLaughlin, P. (2012), 'Ecological modernization in evolutionary perspective', *Organization & Environment*, **25** (2), 178–196.

Mitra, R. and P.M. Buzzanelli (2017), 'Communicative tensions of meaningful work: The case of sustainability practitioners', *Human Relations*, **70** (5), 594–616.

Onkila, T., S. Teerikangas, M. Mäkelä and K. Koistinen (2019), 'Sustainability agency: actors attributes and strategies – a systematic review of CSR literature', Paper presented at Corporate Responsibility Research Conference, 12–13 September, 2019, Tampere, Finland.

O'Rourke, A. (2003), 'A new politics of engagement: Shareholder activism for corporate social responsibility', *Business Strategy and the Environment*, **12** (4), 227–239.

Papoutsi, A. and M.S. Sodhi (2020), 'Does disclosure in sustainability reports indicate actual sustainability performance?', *Journal of Cleaner Production*, **260**, 121049.

Patriotta, G., J. Gond and F. Schultz (2011), 'Maintaining legitimacy: Controversies, orders of worth, and public justifications', *Journal of Management Studies*, **48** (8), 1804–1836.

Peattie, K. and A. Samuel (2018), 'Fairtrade towns as unconventional networks of ethical activism', *Journal of Business Ethics*, **153** (1), 265–282.

Peloza, J. (2009), 'The challenge of measuring financial impacts from investments in corporate social performance', *Journal of Management*, **35** (6), 1518–1541.

Pesch, U. (2015), 'Tracing discursive space: Agency and change in sustainability transitions', *Technological Forecasting and Social Change*, **90**, 379–388.

Pitelis, C.N. (2009), 'The co-evolution of organizational value capture, value creation and sustainable advantage', *Organization Studies*, **30** (10), 1115–1139.

Ritzer, G. (2005), *Sociological Theory* (5th edition), Singapore: McGraw-Hill.

Sarasini, S. and M. Jacob (2014), 'Past, present, or future? Managers' temporal orientations and corporate climate action in the Swedish electricity sector', *Organization & Environment*, **27** (3), 242–262.

Scherer, A.G. and G. Palazzo (2011), 'The new political role of business in a globalized world: A review of a new perspective on CSR and its implications for the firm, governance, and democracy', *Journal of Management Studies*, **48** (4), 899–931.

Shepherd, D.A., H. Patzelt and R.A. Baron (2013), '"I care about nature, but...":
Disengaging values in assessing opportunities that cause harm', *Academy of
Management Journal*, **56** (5), 1251-1273.

Sherwin, S. (2009), 'Relational existence and termination of lives: When embodiment
precludes agency', in Campbell, S., L. Meynell and S. Sherwin (eds.), *Embodiment
and Agency*, University Park: Pennsylvania State University Press.

Sonpar, K., J.M. Handelman and A. Dastmalchian (2009), 'Implementing new institutional logics in pioneering organizations: The burden of justifying ethical appropri-
ateness and trustworthiness', *Journal of Business Ethics*, **90** (3), 345.

Stephenson, J. (2018), 'Sustainability cultures and energy research: An actor
centred interpretation of cultural theory', *Energy Research and Social Science*, **44**,
242-249.

Stones, R. (2005), *Structuration Theory*, Basingstoke: Palgrave Macmillan.

Teerikangas, S., T. Onkila, K. Koistinen and M. Mäkelä (eds.) (2021a), *Research
Handbook on Sustainability Agency*, Cheltenham, UK and Northampton, MA, USA:
Edward Elgar Publishing.

Teerikangas, S., T. Onkila, K. Koistinen and M. Mäkelä (2021b), 'Synthesis and future
research directions', in Teerikangas, S., T. Onkila, K. Koistinen and M. Mäkelä (eds.),
Research Handbook on Sustainability Agency, Cheltenham, UK and Northampton,
MA, USA: Edward Elgar Publishing.

Teerikangas, S., T. Onkila, K. Koistinen, L. Niemi and M. Mäkelä (2018), 'Agency in
sustainability transitions: A closer look at management literatures', A paper presented at the CRR 2018 conference.

Tourish, D. (2014), 'Leadership, more or less? A processual, communication perspective on the role of agency in leadership theory', *Leadership*, **10** (1), 79-98.

van der Vleuten, E. (2018), 'Radical change and deep transitions: Lessons from
Europe's infrastructure transition 1815-2015', *Environmental Innovation and
Societal Transitions*, **32**, 22-32.

Walls, J.L. and P. Berrone (2017), 'The power of one to make a difference: How informal and formal CEO power affect environmental sustainability', *Journal of Business
Ethics*, **145** (2), 293-308.

Whiteman, G. and W.H. Cooper (2000), 'Ecological embeddedness', *Academy of
Management Journal*, **43** (5), 1265-1282.

3 A critical review of the socially responsible consumer

Ning Lu, Phani Kumar Chintakayala, Timothy Devinney, William Young and Ralf Barkemeyer

1. Introduction

Socially responsible consumers are individuals who tend to exhibit intentions or attitudes concerning ethicality or social responsibility when they consume. However, in reality, consumers often fail to translate their intentions or attitudes into actual behaviours, even when they have clearly expressed their commitment to socially responsible consumption (Auger and Devinney, 2007; Young et al., 2010). A number of reasons have been put forward to explain the attitude–behaviour discrepancy, such as measurement bias, value conflicts, moral licensing/decoupling, lack of accessibility, or lack of awareness and engagement (Hui et al., 2009; Merritt et al., 2010; Carrington et al., 2010; Steg et al., 2014; Kumar, 2016; Yamoah et al., 2016). Yet socially responsible consumption is still reportedly growing strong amid this persistent attitude–behaviour gap (Doherty et al., 2015). While it seems that people are growing more conscious about the ethical side of their consumption, ambiguity remains as to whether it is really ethical or socially responsible consciousness that is triggering consumption, given the multiple attributes ascribed to that consumption (Spiteri Cornish, 2013). Example attributes include price, colour, brand, style, size, weight, volume, and material composition.

In this chapter, we intend to show that research into socially responsible consumption will need to account for a wider variety of prosocial behaviour drivers. Behavioural science has long pursued behaviours in relation to ethics by examining underlying behavioural drivers through social experiments. This differs from traditional research on social responsibility, which focused on justice, value, responsibility, or morality in which the scientific methods utilised are largely non-experimental. The chapter will start by reviewing the main behaviour drivers in socially responsible consumption. This offers a means with which to reflect a continuous integration between consciousness and unconsciousness when examining human behaviours. Building on this review, the chapter then discusses future directions in terms of theories and methods.

2.   Behaviour drivers

2.1   Values as a driver

Ethical values are often considered as the primary motivation for socially responsible consumption (Shaw et al., 2005; Schwartz, 2012; Oh and Yoon, 2014). There are several concepts that have been used to capture ethical values involved in socially responsible consumption. These include altruism, collectivism, hedonism, equality, welfare, law obedience, health, and environmental consciousness. Schwartz (2012) noted that value-based behavioural drivers are relatively stable across time and are able to manifest across different situations. This means that individual values can be seen as a common navigating system for people's consumption. However, it is unlikely that everyone shares the same set of values, and competing values can create gaps in socially responsible consumption. Moreover, value-based behavioural drivers seem to rely heavily on consumers' rational decision-making but behavioural economics have consistently shown the dominance of their unconscious side (Kahneman, 2011; Thaler and Sunstein, 2009).

The behavioural mechanism behind prosocial values is that consumers are made to realise their 'civic duty' and moral concerns. In return, they will direct their consumption akin to political voters (Shaw et al., 2006). In a way, the ethical values influence beliefs, these beliefs translate into intentions, and intentions dictate individual consumption behaviour (Shaw et al., 2006; Soper, 2007; Balderjahn et al., 2013; do Paço et al., 2013). A number of studies examining ethical preferences and beliefs towards the purchase intention of socially responsible goods have shown a positive relationship (Ma and Lee, 2012; Haws et al., 2014; Nguyen et al., 2017).

2.2   Attitudes as a driver

The theories most commonly used to explain consumption behaviour mechanisms are the theory of reasoned action (TRA) and the theory of planned behaviour (TPB) (Ajzen, 1985; Ajzen and Fishbein, 1980; Carrington et al., 2010). TRA starts by decomposing the attitude–behaviour relationship into beliefs, attitudes, intentions, and behaviour (Madden et al., 1992). It argues that individual behaviour is conditional on individual intentions while individual intentions are conditional on both individual attitudes and subjective norms (Olson and Zanna, 1993). If an individual expresses his or her desire to commit to socially responsible consumption and if he or she believes that the action fits the social norm, that individual is more likely to match his or her behaviour with a given socially responsible attitude. However, one of the main downsides of TRA is that it fails to take perceived behavioural control into account (Sideridis et al., 1998). TPB is built upon TRA by including the component of individual perceived behaviour control; hence, the theoretical model more comprehensively addresses the attitude–behaviour gap (Sideridis et al., 1998; Madden et al., 1992). There is consistent empirical support for both TRA and TPB (Antonetti and Maklan, 2014; Handayani, 2017). Pagiaslis and Krontalis (2014) show that intentions to use and beliefs about biofuels have a significant positive relationship with the intention to pay a premium price for those fuels. Göçer and Oflaç (2017) explored the factors influencing eco-labelled products in Turkey and found that the perceived environmental knowledge of young consumers had a positive relationship with tendencies to purchase eco-labelled products, with environmental concerns having a mediating effect. A study examining the relationship between environmental attitudes and behaviour shows that the environmental attitudes had a positive relationship with a high degree of collectivism and law obedience, but more importantly, they have a significant impact on the adoption of the environmental behaviour (Leonidou et al., 2010). Another study examined consumer attitudes and willingness to pay a premium for a local and organic cotton shirt (Ha-Brookshire and Norum, 2011). It shows that people with high socially responsible attitudes tend to show a higher willingness to pay a premium for sustainable apparel.

However, a number of studies demonstrate that attitudes do not necessarily get translated into socially responsible consumption, particularly when consumers face abundant ethical prepositions (Uusitalo and Oksanen, 2004; Nittala, 2014). Leonidou et al. (2015) argue that consumer attitudes related to ethical and social responsibility are not free from influence; there are background forces that are often neglected that are in fact responsible for motivation. Uusitalo and Oksanen (2004), examining the link between consumer attitude and intention for socially responsible consumption, found that common problems were a lack of knowledge about actual purchasing behaviour and reliable information about products related to social responsibility. This means that researchers tend to get spurious correlations by measuring consumer attitude and outcome through the same survey or using self-reported measures but never verify actual consumption. These potential biases are discussed as common method bias in behavioural sciences but they appear to be neglected when examining consumer attitude concerning social responsibility and socially responsible consumption (Strizhakova and Coulter, 2013; Tsarenko et al., 2013; He et al., 2016; Nguyen et al., 2017).

In summary, attitude-based behavioural drivers are about intention and willingness. If the gravitation of one's attitude towards socially responsible consumption is strong enough, it is very likely that attitude will translate into actual socially responsible behaviours. However, intention and willingness often fall short of actual behaviours. Research into both self-reported and common method biases have shown that consumers often fail to translate their intentions or attitudes into actual behaviours, even when they have clearly expressed their commitment to socially responsible consumption. The move from TRA to TPB has also signalled the theoretical shift towards a more concrete and formulated nature of intention, such as the intention to implement such actions, and attitude-based behavioural drivers are increasingly accompanied by more volatile measures, such as actual behavioural control and situational factors.

2.3   Situations as a driver

In contrast to conscious information processing, situational drivers are subtle, unconscious, and activating social conformity as well as automatic goal pursuit within the individual (Dijksterhuis et al., 2005; Dahl et al., 2016). Socially responsible consumers are susceptible to situational influences just like every-day consumers (Schultz et al., 1995; Simpson and Radford, 2014). Situational drivers tend to interact across multiple levels and it could be conceptualised as a situational taxonomy ranging from the micro- to the macro-level (Belk, 1975; Bonner, 1985; Nair and Little, 2016; Milfont and Markowitz, 2016).

2.4   Macro-level situations

At a higher level, situational drivers can expand into social, economic, political, technological, temporal, media, or government domains, and be associated with cultural influence and lifestyle (Lee, 2010; Chen and Lobo, 2010). One study found Slovenian consumers more likely to score higher than average on self-reported pro-environmental behaviour than French consumers (Culiberg and Elgaaied-Gambier, 2016). Cultural differences have also consistently been highlighted in other country contexts. Cho and Krasser (2011) found that consumers in Austria had a greater motivation to engage in socially responsible consumerism than consumers in South Korea did. Dermody et al. (2015) also found that, when compared to their UK counterparts, Chinese consumers tended to express more desire for symbolic as well as socially responsible consumption. Morren and Grinstein (2016) conducted a meta-analysis of environmental behaviours; their findings support the 'affluence hypothesis', that is, that people with high socioeconomic status in developed and individualistic countries are more likely to enact environmental behaviours. However, some of the observed cultural differences might also just be due to variances in translation (Erffmeyer et al., 1999).

2.5   Micro-level situations

At the individual level, situational drivers are often discussed in relation to the presence or engagement of others, such as bystander effect, group pressure, and participatory effect (Bagozzi et al., 2000; Garcia et al., 2009; Norton et al., 2012). One review study examining food consumption context shows that a dining group can have a significant impact on the amount of food consumption beside food size, plate shape, lighting, and the layout of food choices (Wilcox et al., 2009; Wansink and Chandon, 2014). Another study examining the impact of the choices made by others found that there was a strong heterogeneity across different consumer groups and suggests that segmentation is the key to dissect the underlying threshold (Wheeler and Berger, 2007). Participatory situations remain a powerful means to increasing the likelihood of socially responsible consumption. One study recommends non-profit sectors to consider introducing a participatory situation for potential donors as it found that consumers were more likely to respond to the socially responsible campaign when they were given the choice to nominate their own sponsored cause (Howie et al., 2018). Such effect is an example of the so-called 'Ikea effect' (Norton et al., 2012) whereby consumers attach a higher value to products created or suggested by them than to other products (Norton et al., 2012).

One way to explore the heterogeneity raised by different situations is to look into the psychological interaction with the situations, such as emotion, goal adoption, and perception (Aarts et al., 2004; Reis and Holmes, 2012; Ertz et al., 2016). Consumers deal with multiple goals each day – shopping, dining, watching television, and they do it either in isolation or as part of a group. Yet they are very likely to adopt or share another person's goal. Sela and Shiv (2009) show that goal-directed behaviour can be activated through subtle environmental cues, such as semantic activation. In socially responsible consumption, Lamberton (2016) describes a goal-sharing form of consumption involved with car sharing, community gardens, and toys. However, Papies (2016) found that environmental cues were more likely to activate short-term hedonic goals than long-term investment goals. Malti and Dys (2018) cited that infants and young children could demonstrate prosocial behaviours towards their close peers who also showed patterns of prosocial behaviour. This reminds us that the mechanism behind social situations causing individuals to become involved with others is likely rooted in human instinct, but it is in need of situational cues to be activated.

In summary, situational drivers remain among the most effective behavioural drivers to alter prosocial behaviours. This idea has proven to be useful when examining how socially responsible consumers might succeed or fail to align their intention with actual behaviour. However, this remains challenging as situational drivers are complex, existing as they do on macro-, and micro-levels. Hence, it can be difficult to account for one level of situational influence in isolation whilst disregarding other levels.

2.6   Signalling as a driver

Socially responsible consumption can be seen as a part of the social signalling process concerning status seeking, identity seeking, and conspicuous conservation (Griskevicius et al., 2010; Ariely et al., 2009; Sexton and Sexton, 2014; Brick et al., 2017). The underlying premise is that a signal is a costly action that alleviates asymmetrical information; that information must be credible in order for the signal to work (Ariely et al., 2009). For example, when Hyundai deployed a sales promotion offering 'America's Best Warranty', that warranty acted as a signal to consumers about the manufacturer's faith in the quality of their cars (Melewar et al., 2007).

Thorstein Veblen (Gino, 2018, p.26), who coined the term 'conspicuous consumption', argued that 'redounded to their glory, and now the middle class was using its newfound wealth to purchase elite status'. This suggests that when a society is making advances in terms of socioeconomic progress, making their middle-class consumers relatively stable, the consumers are expected to look for new ways to seek status and identity, as well as to engage in conspicuous conservation behaviours. Griskevicius et al. (2010) shows that people who own hybrid gas–electric cars, such as the Toyota Prius, do not necessarily score high in environmental conservation. However, a Toyota Prius is expensive, and thus ownership enables the car owners to not only enjoy its fuel-efficient utility, but also to give a higher social status signal to their peers. A similar observation was made by Yan et al. (2010) with regard to the purchase intentions of young consumers towards American Apparel. They found that consumer motivation in young people was heavily linked to perceptions of source credibility, but it failed to reject the possibility of status seeking in young people about their lifestyle traits, such as spending power. Small and Cryder (2016) showed that when philanthropists are given naming rights to buildings and parks, their donations tend to be larger. The reason is that the naming right enables the philanthropists to signal their generosity in public domains, such as newspapers or social media. Puska et al. (2018) found evidence for the signalling effect in organic consumption. When consumers were primed with a desire for high status, they tended to opt for organic rather than non-organic food. Once again, the reason might be that organic food consumption sends out a costly signal about lifestyle status.

In summary, signal-based drivers are one of the least discussed behavioural drivers in the literature on socially responsible consumption, despite the fact that such drivers are helpful when exploring the irrational and unconscious side of human behaviour. The contrarian truth from signal-based drivers is that motivation in socially responsible consumption might not originate from ethics or morality, at least not entirely. This is a reminder that the primary motivator of socially responsible consumption can often be behavioural drivers other than ethicality or social responsibility.

2.7   Narcissism as a driver

One of the personality traits that has attracted growing interest from both the media and consumer behaviour research is narcissism (Meyer and Speakman, 2016). Understanding narcissism has provided not only a means to explain some forms of conspicuous consumption, such as lavish clothes or pet jewellery, but also a new perspective to examine the underlying causal mechanism in socially responsible consumption (Sedikides et al., 2007; Bergman et al., 2014; Naderi and Strutton, 2014; Piff, 2014).

The argument that there has been a rise of narcissistic attitudes among the young population has been made across generations (Stinson et al., 2008; Twenge, 2014). This growth is often associated with the term 'Generation me', which refers to people born in the 1970s, 1980s, and 1990s (Twenge et al., 2008). While narcissism can certainly stimulate a great deal of conspicuous or vanity consumption, some recent evidence has suggested that narcissism can induce some forms of socially responsible consumption (Sedikides et al., 2007; Bergman et al., 2014; Naderi and Strutton, 2014; Piff, 2014). Conventional thinking is that narcissism and prosocial behaviour are two contradictory behavioural artefacts – the former is self-orientated and the latter is community-orientated (Konrath and Tian, 2018). However, it is difficult to classify the underlying motivation for many prosocial behaviours as ethicality, social responsibility, or altruism, simply based on the presence of prosocial behaviours. People with highly developed narcissistic traits are able to enact prosocial behaviours if the behaviour feeds their status-seeking metrics. In other words, socially responsible consumption can simply serve as a means to a narcissist's selfish end when the situation is narcissistic-enabling. Sedikides et al. (2007) suggested that narcissists are more willing to show off and adopt new forms of consumption if those behaviours support their symbolic value and feed their status seeking.

Naderi and Paswan (2016) showed that narcissists tend to process product information differently from non-narcissists, as they value the image of the store more than the price of the product. Narcissists tend to link consumption metrics with a positive self-portraying image. A five-studies experimental research shows how narcissists can declare themselves as environmentalists by inflating their willingness to purchase eco-friendly computers and cars if everyone can see their purchases publicly or the level of self-sacrifice is low (Naderi, 2018). When altruism and consciousness motivate consumer behaviour, it is possible to elicit the effect of individual narcissism. In other words, the success of these social campaigns was driven by the nature of altruism as well as the nature of narcissism. The characteristics of narcissism are often disguised by an altruism-driven appearance (Konrath et al., 2016). This raises the question as to whether the rise in socially responsible consumption is associated with some forms of narcissism. Some recent evidence has indeed suggested that narcissistic traits may motivate consumers to engage in 'green' behaviours (Naderi and Strutton, 2015). By cultivating situational stimuli such as public visibility, people with high narcissistic attitudes are more likely to engage in responsible consumption (Naderi and Strutton, 2014).

Interestingly, most consumer studies examining socially responsible behaviours on the basis of personality traits tend to focus on 'virtuous traits', such as openness, conscientiousness, agreeableness, honesty–humility (Song and Kim, 2018; Engel and Szech, 2020; Riefolo, 2014). Yet 'vice traits' such as narcissism could potentially reveal more insight as socially responsible consumption is often intertwined with some manifestation of norm-violating or status-organising (Anderson and Cowan, 2014; Voyer, 2015). This will open up promising avenues for future research linked to the irrational and unconscious side of responsible consumption and human behaviour.

3.   Future research

One direction in developing future theoretical frameworks is to consider the idea of socially responsible consumption as a complex systems process (Culiberg, 2014; Nair and Little, 2016) – combining the goal mechanism with psychological and situational extensions. It is clear that the main theories in socially responsible consumption all pivot from a pure conscious behaviour approach. Some of the new extensions are focused on various types of value orientation, such as altruism and environmentalism. However, most theories are making improvements in understanding unconscious behavioural drivers, with the most common being situational priming and self-representation. The increasing emphasis of unconscious behavioural drivers shows that consumers do not simply live in a vacuum and their decision process is more malleable than expected (Gruber, 2012). The aim of a theoretical framework is to only take into account those factors that contribute to the decision to enact socially responsible consumption, while assuming that other behavioural drivers remain constant (do Paço et al., 2013). But this is unlikely to reflect the full reality of the process. For example, what causes someone to buy a recycled sports shoe? The simplest answer is that the individual has an environmental concern because the expectation is that socially responsible consumption can only be motivated by one's value orientation or ethical belief. However, there are other conditions under which individuals might opt for socially responsible consumption, beyond being concerned about value orientation or ethical beliefs. Socially responsible consumption is a new emergent phenomenon that is increasingly a topic of discussion. However, it is also a complex system product with ethical, social, psychological, financial, and environmental contributions. The theoretical framework should be flexible enough to accommodate the presence of multiple behavioural drivers as well as simple enough to allow for comparison, statistical modelling, and simulation.

Experimentation remains significantly underutilised in socially responsible consumption research, especially when looking at the effect of motivation malleability on socially responsible consumption. With the rise of big data and mobile devices, experimentation can be easily implemented through social media apps or branded mobile apps. Many consumer researchers are already using social media data to enrich their understanding of socially responsible consumers; however, only a few researchers have used social media platforms as experimental venues to gain further conversations and tease out the underlying behavioural drivers through randomised control trial. Such experimental methods include chat bot research, app notification priming, and in-app experimentation.

Experiments are not only able to examine the interaction between socially responsible goods and consumer segments, but can also uncover hidden consumer insights through a behavioural lens, especially in connection with loyalty card schemes or transactions. Socially responsible consumption is not a zero-sum game, but the devil remains in the vague consensus driven by the consumer surveys. It is extremely risky to bet on a survey finding if the consumers do not often practise what they preach. In addition, mobile experimentation can be done remotely, allowing researchers to reduce self-reporting bias as the participants can remain in a real consumer environment rather than a university setting. Experimentation is an excellent means of providing further validation for researchers as well as policy makers on what really drives individuals towards socially responsible consumption.

4.   Conclusion

The motivation behind the rise in socially responsible consumers is often believed to be growing concerns regarding ethicality or social responsibility. However, long-established evidence of an attitude–behaviour gap indicates otherwise (Carrington et al., 2016): consumers often fail to translate their intentions or attitudes related to ethicality or social responsibility into their actual behaviours, even when they clearly express their commitment to socially responsible consumption. If behaviours are finally aligning with socially responsible intentions or attitudes, what could be the behavioural driver in making them do so? We have argued that the rise in socially responsible consumers is potentially the resultant force of more than one behavioural driver, and more importantly, the idea of socially responsible consumers has not yet been fully explored. Most behavioural drivers have shown their capability or effectiveness in motivating socially responsible consumption. However, no single behavioural driver is able to explain all the variance in observed behaviour. Thus, it is key to understand how these behavioural drivers have been paired and assembled into a theoretical framework and how these theoretical frameworks succeed or fail in capturing the behavioural mechanics of socially responsible consumption. That being the case, the question remains as to whether we can refer to the consumer who is motivated by behavioural drivers other than ethicality or socially responsibility as truly being socially responsible, or whether this label is in fact misleading. This suggests that future consumer research concerning social responsibility should examine the idea of a complex system in which an interconnection across agents, traits, and situations could play a more significant role in driving and explaining socially responsible behaviours.

Bibliography

Aarts, H., Gollwitzer, P.M. and Hassin, R.R., 2004. Goal contagion: Perceiving is for pursuing. *Journal of Personality and Social Psychology*, 87(1), pp.23–37.

Ajzen, I., 1985. From intentions to actions: A theory of planned behavior. Edited by Julius Kuhl and Jürgen Beckmann, in *Action Control* (pp.11–39). Berlin, Heidelberg: Springer.

Ajzen, I. and Fishbein, M., 1980. Theory of action reasoned. *Journal of Experimental Social Psychology*, 6, pp.466–487.

Anderson, C. and Cowan, J., 2014. Personality and status attainment: A micropolitics perspective. Joey T. Cheng, Jessica L. Tracy, Cameron Anderson, in *The Psychology of Social Status* (pp.99–117). New York: Springer.

Antonetti, P. and Maklan, S., 2014. Feelings that make a difference: How guilt and pride convince consumers of the effectiveness of sustainable consumption choices. *Journal of Business Ethics*, 124(1), pp.117–134.

Ariely, D. and Norton, M.I., 2009. Conceptual consumption. *Annual Review of Psychology*, 60, pp.475–499.

Ariely, D., Bracha, A. and Meier, S., 2009. Doing good or doing well? Image motivation and monetary incentives in behaving prosocially. *American Economic Review*, 99(1), pp.544–555.

Auger, P. and Devinney, T.M., 2007. Do what consumers say matter? The misalignment of preferences with unconstrained ethical intentions. *Journal of Business Ethics*, 76(4), pp.361–383.

Azizan, S.A.M. and Suki, N.M., 2017. Consumers' intentions to purchase organic food products. Edited by Thangasamy Esakki, in *Green Marketing and Environmental Responsibility in Modern Corporations* (pp.68–82). Hershey, PA: IGI Global.

Bagozzi, R.P., Wong, N., Abe, S. and Bergami, M., 2000. Cultural and situational contingencies and the theory of reasoned action: Application to fast food restaurant consumption. *Journal of Consumer Psychology*, 9(2), pp.97–106.

Balderjahn, I., Buerke, A., Kirchgeorg, M., Peyer, M., Seegebarth, B. and Wiedmann, K.P., 2013. Consciousness for sustainable consumption: Scale development and new insights in the economic dimension of consumers' sustainability. *AMS Review*, 3(4), pp.181–192.

Belk, R.W., 1975. Situational Variables and Consumer Behavior, *Journal of Consumer Research*, 2(3), pp.157–164.

Bergman, J.Z., Westerman, J.W., Bergman, S.M., Westerman, J. and Daly, J.P., 2014. Narcissism, materialism, and environmental ethics in business students. *Journal of Management Education*, 38(4), pp.489–510.

Bonner, P.G., 1985. Considerations for situational research. Edited by Elizabeth C. Hirschman and Moris B. Holbrook, in *Advances in Consumer Research Volume 12* (pp.368–373). Provo, UT: Association for Consumer Research.

Brick, C., Sherman, D.K. and Kim, H.S., 2017. 'Green to be seen' and 'brown to keep down': Visibility moderates the effect of identity on pro-environmental behavior. *Journal of Environmental Psychology*, 51, pp.226–238.

Carrington, M.J., Neville, B.A. and Whitwell, G.J., 2010. Why ethical consumers don't walk their talk: Towards a framework for understanding the gap between the ethical purchase intentions and actual buying behaviour of ethically minded consumers. *Journal of Business Ethics*, 97(1), pp.139–158.

Carrington, M.J., Zwick, D. and Neville, B., 2016. The ideology of the ethical consumption gap. *Marketing Theory*, 16(1), pp.21-38.

Chen, J. and Lobo, A., 2010. An Exploratory Study Investigating the Dimensions Influencing Consumers' Purchase Intentions relating to Organic Food in Urban China. *Çin'de Organik Gıda ile ilgili Tüketicilerin Satın Alma Niyetini Etkileyen Boyutlara*. Paper, https://researchbank.swinburne.edu.au/file/d487022f-d2ca-4b0c-b548-62709d7d4083/1/PDF%20%28Published%20version%29.pdf.

Cho, S. and Krasser, A.H., 2011. What makes us care? The impact of cultural values, individual factors, and attention to media content on motivation for ethical consumerism. *International Social Science Review*, 86(1/2), pp.3-23.

Culiberg, B., 2014. Towards an understanding of consumer recycling from an ethical perspective. *International Journal of Consumer Studies*, 38(1), pp.90-97.

Culiberg, B. and Elgaaied-Gambier, L., 2016. Going green to fit in: Understanding the impact of social norms on pro-environmental behaviour, a cross-cultural approach. *International Journal of Consumer Studies*, 40(2), pp.179-185.

Dahl, A.A., Hales, S.B. and Turner-McGrievy, G.M., 2016. Integrating social media into weight loss interventions. *Current Opinion in Psychology*, 9, pp.11-15.

De Groot, J.I. and Steg, L., 2009a. Mean or green: Which values can promote stable pro-environmental behavior? *Conservation Letters*, 2(2), pp.61-66.

De Groot, J.I. and Steg, L., 2009b. Morality and prosocial behavior: The role of awareness, responsibility, and norms in the norm activation model. *The Journal of Social Psychology*, 149(4), pp.425-449.

Dermody, J., Hanmer-Lloyd, S., Koenig-Lewis, N. and Zhao, A.L., 2015. Advancing sustainable consumption in the UK and China: The mediating effect of pro-environmental self-identity. *Journal of Marketing Management*, 31(13-14), pp.1472-1502.

Dijksterhuis, A., Smith, P.K., Van Baaren, R.B. and Wigboldus, D.H., 2005. The unconscious consumer: Effects of environment on consumer behavior. *Journal of Consumer Psychology*, 15(3), pp.193-202.

do Paço, A., Alves, H., Shiel, C. and Filho, W.L., 2013. Development of a green consumer behaviour model. *International Journal of Consumer Studies*, 37(4), pp.414-421.

Doherty, B., Bezençon, V. and Balineau, G., 2015. Fairtrade International and the European market. Edited by Laura T. Raynolds and Elizabeth A. Bennett, in *Handbook of Research on Fair Trade* (pp.316-332). Cheltenham, UK and Northampton, MA, USA: Edward Elgar Publishing.

Engel, J. and Szech, N., 2020. A little good is good enough: Ethical consumption, cheap excuses, and moral self-licensing. *Plos One*, 15(1), art.e0227036.

Erffmeyer, R.C., Keillor, B.D. and LeClair, D.T., 1999. An empirical investigation of Japanese consumer ethics. *Journal of Business Ethics*, 18(1), pp.35-50.

Ertz, M., Karakas, F. and Sarigöllü, E., 2016. Exploring pro-environmental behaviors of consumers: An analysis of contextual factors, attitude, and behaviors. *Journal of Business Research*, 69(10), pp.3971-3980.

Garcia, S.M., Weaver, K., Darley, J.M. and Spence, B.T., 2009. Dual effects of implicit bystanders: Inhibiting vs. facilitating helping behavior. *Journal of Consumer Psychology*, 19(2), pp.215-224.

Gino, F., 2018. *Rebel Talent: Why it Pays to Break The Rules at Work and in Life*. London: Pan Macmillan.

Göçer, A. and Oflaç, B.S., 2017. Understanding young consumers' tendencies regarding eco-labelled products. *Asia Pacific Journal of Marketing and Logistics*, 29(1), pp.80-97.

Griskevicius, V., Tybur, J.M. and Van den Bergh, B., 2010. Going green to be seen: Status, reputation, and conspicuous conservation. *Journal of Personality and Social Psychology*, 98(3), pp.392–404.

Gruber, V., 2012. Sustainable consumption decisions: An examination of consumer cognition and behavior. PhD thesis, WU Vienna.

Ha-Brookshire, J.E. and Norum, P.S., 2011. Willingness to pay for socially responsible products: Case of cotton apparel. *Journal of Consumer Marketing*, 28(5), pp.344-353.

Handayani, W., 2017. Green consumerism: An eco-friendly behaviour form through the green product consumption and green marketing. *Sinergi: Jurnal Ilmiah Ilmu Manajemen*, 7(2), pp.25-29.

Haws, K.L., Winterich, K.P. and Naylor, R.W., 2014. Seeing the world through GREEN-tinted glasses: Green consumption values and responses to environmentally friendly products. *Journal of Consumer Psychology*, 24(3), pp.336-354.

He, A.Z., Cai, T., Deng, T.X. and Li, X., 2016. Factors affecting non-green consumer behaviour: An exploratory study among Chinese consumers. *International Journal of Consumer Studies*, 40(3), pp.345-356.

Howie, K.M., Yang, L., Vitell, S.J., Bush, V. and Vorhies, D., 2018. Consumer participation in cause-related marketing: An examination of effort demands and defensive denial. *Journal of Business Ethics*, 147(3), pp.679-692.

Huang, R. and Chen, D., 2015. Does environmental information disclosure benefit waste discharge reduction? Evidence from China. *Journal of Business Ethics*, 129(3), pp.535-552.

Hui, S.K., Bradlow, E.T. and Fader, P.S., 2009. Testing behavioral hypotheses using an integrated model of grocery store shopping path and purchase behavior. *Journal of Consumer Research*, 36(3), pp.478-493.

Kahn, M.E., 2007. Do greens drive Hummers or hybrids? Environmental ideology as a determinant of consumer choice. *Journal of Environmental Economics and Management*, 54(2), pp.129-145.

Kahneman, D., 2011. *Thinking, Fast and Slow*. London: Macmillan.

Konrath, S. and Tian, Y., 2018. Narcissism and prosocial behavior. Edited by Anthony D. Hermann, Amy B. Brunell, Joshua D. Foster, in *Handbook of Trait Narcissism* (pp.371-378). Cham: Springer.

Konrath, S., Ho, M.H. and Zarins, S., 2016. The strategic helper: Narcissism and prosocial motives and behaviors. *Current Psychology*, 35(2), pp.182-194.

Kumar, P., 2016. State of green marketing research over 25 years (1990-2014): Literature survey and classification. *Marketing Intelligence & Planning*, 34(1), pp.137-158.

Lamberton, C., 2016. Collaborative consumption: A goal-based framework. *Current Opinion in Psychology*, 10, pp.55-59.

Lee, K., 2010. The green purchase behavior of Hong Kong young consumers: The role of peer influence, local environmental involvement, and concrete environmental knowledge. *Journal of International Consumer Marketing*, 23(1), pp.21-44.

Leonidou, L.C., Leonidou, C.N. and Kvasova, O., 2010. Antecedents and outcomes of consumer environmentally friendly attitudes and behaviour. *Journal of Marketing Management*, 26(13-14), pp.1319-1344.

Leonidou, L.C., Coudounaris, D.N., Kvasova, O. and Christodoulides, P., 2015. Drivers and outcomes of green tourist attitudes and behavior: Sociodemographic moderating effects. *Psychology & Marketing*, 32(6), pp.635-650.

Ma, Y.J. and Lee, H.H., 2012. Understanding consumption behaviours for fair trade non-food products: Focusing on self-transcendence and openness to change values. *International Journal of Consumer Studies*, 36(6), pp.622-634.

Madden, T.J., Ellen, P.S. and Ajzen, I., 1992. A comparison of the theory of planned behavior and the theory of reasoned action. *Personality and Social Psychology Bulletin*, 18(1), pp.3-9.

Malti, T. and Dys, S.P., 2018. From being nice to being kind: Development of prosocial behaviors. *Current Opinion in Psychology*, 20, pp.45-49.

Melewar, T.C., Small, J., Andrews, M. and Kim, D., 2007. Revitalising suffering multinational brands: An empirical study. *International Marketing Review*, 24(3), pp.350-372.

Merritt, A.C., Effron, D.A. and Monin, B., 2010. Moral self-licensing: When being good frees us to be bad. *Social and Personality Psychology Compass*, 4(5), pp.344-357.

Meyer, H.K. and Speakman, B., 2016. Journalists and mobile: Melding social media and social capital. Edited by Xiaoge Xu, in *Handbook of Research on Human Social Interaction in the Age of Mobile Devices* (pp.200-219). Hershey, PA: IGI Global.

Milfont, T.L. and Markowitz, E., 2016. Sustainable consumer behavior: A multilevel perspective. *Current Opinion in Psychology*, 10, pp.112-117.

Morren, M. and Grinstein, A., 2016. Explaining environmental behavior across borders: A meta-analysis. *Journal of Environmental Psychology*, 47, pp.91-106.

Naderi, I., 2018. I'm nice, therefore I go green: An investigation of pro-environmentalism in communal narcissists. *Journal of Environmental Psychology*, 59, pp.54-64.

Naderi, I. and Paswan, A.K., 2016. Narcissistic consumers in retail settings. *Journal of Consumer Marketing*, 33(5), pp.376-386.

Naderi, I. and Strutton, D., 2014. Can normal narcissism be managed to promote green product purchases? Investigating a counterintuitive proposition. *Journal of Applied Social Psychology*, 44(5), pp.375-391.

Naderi, I. and Strutton, D., 2015. I support sustainability but only when doing so reflects fabulously on me: Can green narcissists be cultivated? *Journal of Macromarketing*, 35(1), pp.70-83.

Nair, S.R. and Little, V.J., 2016. Context, culture and green consumption: A new framework. *Journal of International Consumer Marketing*, 28(3), pp.169-184.

Nguyen, T.N., Lobo, A. and Greenland, S., 2017. Energy efficient household appliances in emerging markets: The influence of consumers' values and knowledge on their attitudes and purchase behaviour. *International Journal of Consumer Studies*, 41(2), pp.167-177.

Nguyen, T.N., Lobo, A. and Nguyen, B.K., 2018. Young consumers' green purchase behaviour in an emerging market. *Journal of Strategic Marketing*, 26(7), pp.583-600.

Nittala, R., 2014. Green consumer behavior of the educated segment in India. *Journal of International Consumer Marketing*, 26(2), pp.138-152.

Norton, M.I., Mochon, D. and Ariely, D., 2012. The IKEA effect: When labor leads to love. *Journal of Consumer Psychology*, 22(3), pp.453-460.

Oh, J.C. and Yoon, S.J., 2014. Theory-based approach to factors affecting ethical consumption. *International Journal of Consumer Studies*, 38(3), pp.278-288.

Olson, J.M. and Zanna, M.P., 1993. Attitudes and attitude change. *Annual Review of Psychology*, 44(1), pp.117-154.

Pagiaslis, A. and Krontalis, A.K., 2014. Green consumption behavior antecedents: Environmental concern, knowledge, and beliefs. *Psychology & Marketing*, 31(5), pp.335-348.

Papies, E.K., 2016. Goal priming as a situated intervention tool. *Current Opinion in Psychology*, 12, pp.12-16.

Petty, R.E., Wegener, D.T. and Fabrigar, L.R., 1997. Attitudes and attitude change. *Annual Review of Psychology*, 48(1), pp.609-647.

Piff, P.K., 2014. Wealth and the inflated self: Class, entitlement, and narcissism. *Personality and Social Psychology Bulletin*, 40(1), pp.34-43.

Puska, P., Kurki, S., Lähdesmäki, M., Siltaoja, M. and Luomala, H., 2018. Sweet taste of prosocial status signaling: When eating organic foods makes you happy and hopeful. *Appetite*, 121, pp.348-359.

Reis, H.T. and Holmes, J.G., 2012. Perspectives on the situation. Edited by Kay Deaux and Mark Snyder, in *The Oxford Handbook of Personality and Social Psychology* (pp.64–92). New York: Oxford University Press.

Riefolo, G., 2014. Personality traits and prosocial behavior: How subjective characteristics may impact on consumption habits. Thesis, Libera Università Internazionale degli Studi Sociali Guido Carli.

Rubenstein, J.C., 2016. The lessons of effective altruism. *Ethics & International Affairs*, 30(4), pp.511-526.

Schultz, P.W., Oskamp, S. and Mainieri, T., 1995. Who recycles and when? A review of personal and situational factors. *Journal of Environmental Psychology*, 15(2), pp.105-121.

Schwartz, S.H., 2012. An overview of the Schwartz theory of basic values. *Online Readings in Psychology and Culture*, 2(1), art.11.

Sedikides, C., Gregg, A.P., Cisek, S. and Hart, C.M., 2007. The I that buys: Narcissists as consumers. *Journal of Consumer Psychology*, 17(4), pp.254-257.

Sela, A. and Shiv, B., 2009. Unraveling priming: When does the same prime activate a goal versus a trait? *Journal of Consumer Research*, 36(3), pp.418-433.

Sexton, S.E. and Sexton, A.L., 2014. Conspicuous conservation: The Prius halo and willingness to pay for environmental bona fides. *Journal of Environmental Economics and Management*, 67(3), pp.303-317.

Shaw, D., Newholm, T. and Dickinson, R., 2006. Consumption as voting: An exploration of consumer empowerment. *European Journal of Marketing*, 40(9/10), pp.1049-1067.

Shaw, D., Grehan, E., Shiu, E., Hassan, L. and Thomson, J., 2005. An exploration of values in ethical consumer decision making. *Journal of Consumer Behaviour: An International Research Review*, 4(3), pp.185-200.

Sideridis, G.D., Kaissidis, A. and Padeliadu, S., 1998. Comparison of the theories of reasoned action and planned behaviour. *British Journal of Educational Psychology*, 68(4), pp.563-580.

Simpson, B.J. and Radford, S.K., 2014. Situational variables and sustainability in multi-attribute decision-making. *European Journal of Marketing*, 48(5/6), pp.1046-1069.

Small, D.A. and Cryder, C., 2016. Prosocial consumer behavior. *Current Opinion in Psychology*, 10, pp.107-111.

Song, S.Y. and Kim, Y.K., 2018. Theory of virtue ethics: Do consumers' good traits predict their socially responsible consumption? *Journal of Business Ethics*, 152(4), pp.1159-1175.

Soper, K., 2007. Re-thinking the Good Life: The citizenship dimension of consumer disaffection with consumerism. *Journal of Consumer Culture*, 7(2), pp.205-229.

Spiteri Cornish, L., 2013. Ethical consumption or consumption of ethical products? An exploratory analysis of motivations behind the purchase of ethical products. *ACR North American Advances*, 41, 337–341.

Steg, L., Bolderdijk, J.W., Keizer, K. and Perlaviciute, G., 2014. An integrated framework for encouraging pro-environmental behaviour: The role of values, situational factors and goals. *Journal of Environmental Psychology*, 38, pp.104-115.

Stinson, F.S., Dawson, D.A., Goldstein, R.B., Chou, S.P., Huang, B., Smith, S.M., Ruan, W.J., Pulay, A.J., Saha, T.D., Pickering, R.P. and Grant, B.F., 2008. Prevalence, correlates, disability, and comorbidity of DSM-IV narcissistic personality disorder: Results from the wave 2 national epidemiologic survey on alcohol and related conditions. *The Journal of Clinical Psychiatry*, 69(7), pp.1033–1045.

Strizhakova, Y. and Coulter, R.A., 2013. The 'green' side of materialism in emerging BRIC and developed markets: The moderating role of global cultural identity. *International Journal of Research in Marketing*, 30(1), pp.69–82.

Thaler, R.H. and Sunstein, C.R., 2009. *Nudge: Improving Decisions about Health, Wealth, and Happiness*. London: Penguin.

Tsarenko, Y., Ferraro, C., Sands, S. and McLeod, C., 2013. Environmentally conscious consumption: The role of retailers and peers as external influences. *Journal of Retailing and Consumer Services*, 20(3), pp.302–310.

Twenge, J.M., 2014. *Generation Me – Revised and Updated: Why Today's Young Americans are More Confident, Assertive, Entitled – and More Miserable than Ever Before*. New York: Simon & Schuster.

Twenge, J.M., Konrath, S., Foster, J.D., Keith Campbell, W. and Bushman, B.J., 2008. Egos inflating over time: A cross-temporal meta-analysis of the Narcissistic Personality Inventory. *Journal of Personality*, 76(4), pp.875–902.

Uusitalo, O. and Oksanen, R., 2004. Ethical consumerism: A view from Finland. *International Journal of Consumer Studies*, 28(3), pp.214–221.

van der Linden, S., 2017. The nature of viral altruism and how to make it stick. *Nature Human Behaviour*, 1(3), pp.1–4.

Veblen, T., 2005. *Conspicuous Consumption* (Vol. 38). London: Penguin.

Voyer, P., 2015. Consumer proclivity for sustainable consumption: A social normative approach. *Advances in Consumer Research*, 43, 421–427.

Wansink, B. and Chandon, P., 2014. Slim by design: Redirecting the accidental drivers of mindless overeating. *Journal of Consumer Psychology*, 24(3), pp.413–431.

Wheeler, S.C. and Berger, J., 2007. When the same prime leads to different effects. *Journal of Consumer Research*, 34(3), pp.357–368.

Wilcox, K., Vallen, B., Block, L. and Fitzsimons, G.J., 2009. Vicarious goal fulfillment: When the mere presence of a healthy option leads to an ironically indulgent decision. *Journal of Consumer Research*, 36(3), pp.380–393.

Yamoah, F.A., Duffy, R., Petrovici, D. and Fearne, A., 2016. Towards a framework for understanding fairtrade purchase intention in the mainstream environment of supermarkets. *Journal of Business Ethics*, 136(1), pp.181–197.

Yan, R.N., Ogle, J.P. and Hyllegard, K.H., 2010. The impact of message appeal and message source on Gen Y consumers' attitudes and purchase intentions toward American Apparel. *Journal of Marketing Communications*, 16(4), pp.203–224.

Young, W., Hwang, K., McDonald, S. and Oates, C.J., 2010. Sustainable consumption: Green consumer behaviour when purchasing products. *Sustainable Development*, 18(1), pp.20–31.

4 Examining both organisational environmental sustainability & organisational resilience: sketching an initial framework

Kerrie L. Unsworth and Rebecca Pieniazek

1.   Introduction

The terms "sustainable" and "resilient" both imply long-lasting phenomena; both have been used to refer to organisations and the environment across various levels of analysis and both are incredibly important for organisations during the climate emergency. In this chapter, we identify the similarities and the differences between organisational environmental sustainability and organisational resilience and, in doing so, argue that they are unique but related constructs that, in combination, can be used to represent particular categories of organisations. More specifically, we delineate four profiles of exemplar organisations: (1) short-term players (low on both organisational environmental sustainability and resilience); (2) survivors (low on organisational environmental sustainability but high on resilience); (3) self-sacrificers (high on organisational environmental sustainability but low on resilience); and (4) resilient environmentalists (high on both organisational environmental sustainability and resilience). These four profiles represent the beginning of a multi-level framework for future research to investigate and unpack further.

This book provides a resource for people interested in many aspects of sustainability; but in this chapter we focus specifically on the environmental component. To date, the academic discourse has generally not considered the distinctiveness nor interplay between organisational resilience (OR) and organisational environmental sustainability (OES) (albeit with a few exceptions, e.g., Espiner et al., 2017). Thus, we investigate their similarities and differences and examine how they might operate together in an organisation, portrayed through four combination profiles of Organisational Environmental Sustainability and Resilience (OESR). In sum, we argue that neither OR nor OES is a sub-type of the other, but that mechanisms relating to enabling or hindering one or the other can have an influence on achieving the other. To begin, however, we look to understand each construct in more detail.

2.   What is organisational resilience? What is organisational environmental sustainability?

Trying to answer the twin questions of "what is OR?" and "what is OES?" is not as simple as one might expect. When considering the former question, many have conceptualised OR as a strategic and behavioural organisational capability to deal with disruptions (Lengnick-Hall et al., 2011). Hollnagel (2010), for example, defines OR as an organisation's propensity to learn, anticipate, monitor and respond, with more complex capability lists covering both proactive and reactive capabilities (Pieniazek, 2017). Others define OR based on outcomes (e.g., Limnios et al., 2014; Ortiz-de-Mandojana & Bansal, 2016), whether that be maintaining the status quo or restoration (i.e., bounce-back) (e.g., Hamel & Välikangas, 2003), or adaptation following disruption (Parsons, 2010). Arguably, organisations use capabilities (as those described above) to achieve these various states, and hence the capability and outcome perspectives are not mutually exclusive (Duchek, 2020).

Adding further to the confusion is the fact that OR is often muddled with employees' personal resilience (e.g., Fisher et al., 2019). This is also defined as "the ability to bounce back or recover from stress" (Smith et al., 2008, p.194) albeit through personal capabilities akin to either personality traits or skills (e.g., Pipe et al., 2012; Windle et al., 2011). This issue of resilience being proposed to have the same meaning at both the organisational and individual level is becoming more apparent (e.g., West et al., 2009) and a number of scholars have highlighted the problems with this approach (e.g., Carmeli et al., 2013; Hartmann et al., 2020). On the one hand, we agree that OR is substantively different to personal resilience (Britt et al., 2016; Kossek & Perrigino, 2016), on the other, we are sympathetic to the view that employees, and their personal attributes including their personal resilience, have a role in performing OR.

Thus, given the entanglement amongst the definitions, we developed a definition of OR that is based not solely on capabilities, nor organisational outcomes, nor individual resiliencies but instead on a Strategy as Practice and Process

(SAPP) perspective (Burgelman et al., 2018; Kouamé & Langley, 2018). In doing so, we incorporate multiple levels of organisational action to define OR as the performance of activities [intended to enable the organisation to deal with, and bounce back from disruptions] at the meso level by actors in the organisation and shaped by the organisation's practices and context. Similar to a process perspective (Williams et al., 2017) it acknowledges that capabilities and procedures must be practised and it is in the praxis itself that OR emerges.

Now that we have a little more clarity on OR, we can begin to think about the latter question, namely, "What is OES?". Like OR, there are many definitions of OES (see e.g., Orlitzky et al., 2011). Some define it as an outcome (e.g., Dilchert & Ones, 2021), some as a set of dynamic capabilities (e.g., Strauss et al., 2017), and others focus on the employee's own environmental actions (see Unsworth et al., 2021). As with resilience, we believe that these are all distinct facets but do not comprise OES in, and of, itself. Instead, we follow the SAPP approach and suggest that OES is *the meso-level performance of actions intended to aid (or at least minimise harm) to the environment by organisational actors, including employees, top management, board members and so forth, within the constraints and supports of organisational practices.*

3. Comparing OES and OR

3.1 Similarities between OES and OR

As can be seen from the previous discussion both OES and OR have multiple definitions in the existing literature. But beyond this, how do they compare with each other? First, OR and OES are similar in that both phenomena are intended to deal, most often, with events that occur externally, for example, floods, climate change, cyber-hacking and so on, by taking action internally within the organisation and/or capitalising on strong external relationships (Johnson & Elliott, 2011; Powley et al., 2017). Furthermore, we can see that OR and OES are inherently linked in this singular purpose. There is a need for OES because our planet is degrading due to human actions; this degradation has increased turbulent natural environments, such as floods and hurricanes and so on, which are some of the very disturbances against which OR aims to protect the organisation.

Second, as a driver of taking such actions, both OES and OR denote the maintenance of something. In the case of OR it is the maintenance of the organisation's operations, functions and/or services, whereas OES aims to maintain the environment (Espiner et al., 2017). This commonality between OES and OR creates much confusion both in academe and in practice because the word "sustainability" is used to refer to the continual maintenance of any or all phenomena or entities, yet it is often used as a shorthand for the maintenance only of the natural world. Thus, OR could actually be called organisational sustainability given that it is focused on the maintenance and ongoing viability (i.e., sustenance) of the organisation. Given that, as noted throughout this book, the term "sustainability" refers to organisational outcomes beyond the environment, we therefore recommend that scholars carefully distinguish between general sustainability (such as triple bottom line approaches; Elkington, 1994) and environmental sustainability in order to reduce this conflation.

Third, scholars emphasise the importance of adapting and learning in the performance of both OR and OES (Larsson et al., 2016). The occurrence of a disruption, or awareness of a potential disruption on the horizon, triggers the temporary undoing and altering of formal practices and relational structures (Powley, 2009; Weick et al., 1999) requiring adaptation. In the field of OR, this is often seen to trigger the temporal space for organisation members to renew OR-relevant relationships (Powley, 2009) as well as enabling the performance of adaptive and dynamic capabilities (e.g., Lengnick-Hall et al., 2011). In the field of OES, the role of adapting and learning in response to external threats has been noted but less well elucidated. On the one hand, "adapting" to climate change is a phenomenon in and of itself, but bears more resemblance to OR than to OES; that is, climate change adaptation is generally seen as a way to protect the organisation and not to protect the environment. On the other hand, recent work empirically demonstrates the importance of learning processes in the development of OES (Battistella et al., n.d.).

Moreover, both OES and OR involve, to some degree, repurposing existing capabilities to different needs when there is a need to protect. Sutcliffe and Vogus (2003) suggest that OR requires a repertoire of capabilities that can be repurposed to fit whatever need emerges. OES scholars typically focus on dynamic capabilities and recommend a more baseline set of proactive capabilities that can be repurposed to enable the organisation to "identify new means for achieving specified ends" (Strauss et al., 2017, p.1344). In other words, both OR and OES recognise the importance of having actors and structures that are able to adapt their existing capabilities to proactively serve the needs of the organisation and the environment, respectively.

3.2   Differences between OR and OES

Thus, there are a lot of similarities between OES and OR yet, they are obviously different in at least one way – they have a different focus of protection. OR is focused on protecting the organisation while OES is focused on protecting the ecological environment. However, when we look to the OR literature, we see that many OR scholars focus on OR towards a particular object, for example supply chain resilience (e.g., Carvalho & Cruz-Machado, 2011) so could there be an argument that "sustainable resilience" is simply the performance of resilience in order to protect environmental sustainability (e.g., Augusta et al., 2017)?

We don't think so. A second difference between OR and OES is their unintended consequences. OR is a drive to protect operations, and unless the organisation is completely reliant on those operations being environmentally friendly (e.g., in a highly scrutinised organisation with an activist market-base), OR will not necessarily be performed in line with best practice for the environment. Similarly, focusing purely on OES may put the organisation at risk due to high costs or lack of redundancy within the system. Thus, sustainable resilience (as an amalgamation of OR and OES) is likely the purview of only a very few organisations who are able to counter the unintended consequences of both OR and OES. While we might be able to determine a clear definition of "high sustainable resilience", there are many dimensions and aspects behind "low sustainable resilience" which make it impossible to define with any precision and which therefore preclude its existence.

Finally, OR has a much narrower focus than OES. At lower levels of analysis, teams or departments within an organisation are focused on their own resilience, even though they could help or hinder each other (Kahn et al., 2018) by distributing or holding back resources (Vogus & Sutcliffe, 2007); organisations have a similar narrow focus on their own resilience. This creates a spotlight on the most important factors related to the ongoing viability and maintenance of the organisational functioning (Lee et al., 2013), and the brightness and intensity of this focus drives motivation to ensure these few factors are dealt with (Unsworth et al., 2014). In contrast, OES creates a diffuse light across the broad range of issues related to the degradation of our planet. At an individual level, this has often been identified as a cause behind a lack of climate action (Kollmuss & Agyeman, 2002), and cognitive psychology has identified the underlying mechanism, called equifinality, for this decrease in motivation when there are many alternative routes to achieving the goal (Kruglanski et al., 2011). We argue that this will likely also be the case at the organisational level,

Table 4.1   Summary of similarities and differences between OR and
OES

	Org Resilience	Org Env. Sustainability
Level of Analysis	Meso-level	Meso-level
Attentional Orientation	External	External
Timeframe	Medium-long term	Medium-long term
Aim	Continuance / maintenance	Continuance / maintenance
Performances	Adaptation & learning	Adaptation & learning
Mechanisms	Repurposing capabilities	Repurposing capabilities
Focus of Protection	Organisation	Environment
Breadth of Threat Focus	Narrow	Broad

and thus, the collective motivations, emotions and processes involved in OES
will necessarily be different to those in OR.

4.   Profiles of organisational environmental sustainability and resilience

So how would OR and OES operate, as separate but related constructs, in the
same organisation? Unfortunately, little research has examined this question
and what has been done has used case studies that exemplify extreme cases.
There are, of course, interdependencies between OR and OES. Given that
scholars have already hinted at the dark sides of OR (Limnios et al., 2014),
and that OR is aimed solely at protecting an organisation's operations, functions and services (van der Vegt et al., 2015), OR could facilitate or hinder
OES depending on whether these operations were harmful or helpful to the
environment. Similarly, it could be that the broad focus of OES could lead
to improved OR via increased external monitoring, or to decreased OR via
diminution of organisational resources. We simply do not know. Furthermore,
the inherent complexity of each of these constructs and their different breadth
of threat foci suggests that integrating OES and OR is not simply a matter
of aligning the protection focus. Instead, as we show below, there are many
nuances involved when we consider the implications of the profiles that
emerge from combining the two.

Table 4.2   Four profiles of organisational environmental sustainability and resilience (OESR)

	Low Sustainability	High Sustainability
Low Resilience	**Short-term players** • Low external orientation • Short-term orientation	**Self-sacrificers** • External & long-term orientation (environmental threats) • Environmental protection focus prioritised • Broad, diffuse threat focus
High Resilience	**Survivors** • External & long-term orientation (organisational threats) • Organisational protection focus prioritised • Narrow, concentrated threat focus	**Resilient environmentalists** • External & long-term orientation • Organisational and environmental protection balanced for synergy • Equilibrium between narrow and broad threat foci

We therefore begin by imagining a matrix that separates organisations into four quadrants, depending on whether they are high or low on OR, and high or low on OES (see Table 4.2). We recognise that such binary distinctions are crude at best, however the four quadrant profiles act as illustrative exemplars of OESR. We then overlay this matrix with the overlapping or differentiating principles (as summarised in Table 4.1), to develop four OESR profiles.

4.1   Short-term players – low OES and low OR

The first profile of OESR that emerges is what we term the "short-term players". These organisations have neither the external orientation nor the long-term timeframe required of both OES and OR. They may have strategically chosen to be short-term players by using start-up businesses to capitalise on short-term market opportunities. On the other hand, they may also have fallen into this category by not considering the OR processes they require, or the OES ecological resources they are consuming, such as energy or water efficiency. In the latter case, the organisations are likely to fail to meet their objectives in the longer term and will likely harm the environment in the process.

4.2   Survivors – low OES and high OR

Our second profile of OESR can be termed "survivors" – their high levels of OR mean that they have a long-term focus, an external awareness of potential disruptions and the processes and mechanisms required to ensure the continuance of the organisation. However, in this quadrant there is perceived to be a trade-off between OR and OES (see e.g., Larsson et al., 2016); for example, by including redundancy of material resources that is useful for OR (Saurin & Werle, 2017; Vogus & Sutcliffe, 2007) but can be harmful for OES. The low priority of OES means that organisations in this quadrant are focusing on how the wider environment will affect organisational maintenance and not on how the organisation might be affecting the maintenance of the wider environment.

The high level of OR performed by Survivors ensures they are aware of, and continually monitor, potential environmental disruptions. Hence, their OES will be focused on elements that might cause such a disruption and are therefore likely to be responsive to environmental regulations and aware of threats to their customer base, reputation and supply chains that might be generated from environmental causes.

There are many examples of companies and industries that fit the Survivors quadrant. Indeed, this has arguably been the template for those in the capitalist-based economy until very recently. For example, Espiner and colleagues (2017) examine the tourism industry and show that, except in a rare "mature state" of sustainability where OES and OR overlap completely, most businesses work towards OR rather than OES. They demonstrate this even in industries that ostensibly aim to be sustainable, such as the skiing industry in New Zealand (Hopkins, 2014, 2015) and nature-based travel (Espiner & Becken, 2014).

We therefore propose that Survivors prioritise the protection of the organisation, even if they are also performing some OES practices. In this way, Survivors view OES as sequentially related to OR – managing OES is a means for achieving OR. Indeed, a number of scholars have highlighted the benefits of OES practices outside of sustainability outcomes. For example, Huang et al. (2020) shows that intentionally deciding to undertake even small amounts of OES means that the business needs to find ways of achieving it; because one of these ways is to build resources, OR is consequently enhanced. Similarly, as noted by Avery and Bergsteiner (2011) and Larsson et al. (2016), sustainable business will invariably help organisational performance. However, we reiterate that Survivors will always prioritise OR over OES – as Mari et al. (2014, p.6666) suggest, "Supply chain systems drop their sustainability objectives while coping with these unexpected disruptions."

4.3   Self-sacrificers – high OES and low OR

Organisational profiles in the third quadrant are focused on OES but not on OR. In this sense, they have a long-term focus and an external orientation as well as the ability to repurpose organisational capabilities to ensure the continuance of the earth. However, the low OR means that little attention is paid to threats and disruptions that emerge from non-environmental sources and the organisation may end up sacrificing itself in search of the greater good.

We have found very little research that has examined Self-Sacrificers. Yet we know that it exists because in our own recent research into social entrepreneurs we found some who fell into a vicious circle of "social suction" – these organisations were so focused on the social aims of their business that they forgot to consider the implications to their business's viability. This is likely a new phenomenon not yet emergent in the academic literature; however we believe that it will likely become more prevalent with the increasing numbers of B-corps and social enterprises.

4.4   Resilient environmentalists – high OES and high OR

The final quadrant is the one which is advocated in the organisational and environmental sustainability field. These organisations use their capability endowments and interact with their surroundings to positively adapt to and improve the functioning of both the organisation and the earth. Resilient environmentalists will perform OR to cope with both environmental and non-environmental threats and disruptions. Moreover, they will be proactively searching for environmentally friendly responses that enable the maintenance of the organisation. Thus, unlike Survivors, they will not simply respond to regulations and enforced mitigation but will identify environmental opportunities to withstand disruption in an environmentally friendly manner, such as the green recovery following the coronavirus pandemic.

Given the hallowed nature of this profile, it attracts most of the extant research. One strand of the literature identifies the necessary capabilities, which has resulted in lists of combined OES and OR attributes. For example, Fiksel (2003) and Augusta et al. (2017) identified four categories of dynamic capabilities required for resilient environmentalism: environmental business strategies, adaptability that leads to learning, eco-efficiency, and organisational cohesion based on shared eco-values. Thus, it appears to be a kind of adaptive OR (e.g., McCarthy et al., 2017) that is underpinned by environmental values.

An alternative to positioning Resilient Environmentalism as OR with an environmental heart is to conceptualise resilience as a multi-dimensional construct. In this we follow Carmeli et al. (2019) who suggested three dimensions to OR: financial, environmental and social. We propose that their sustainability-oriented organisations which are interested in all three dimensions of resilience align with Resilient Environmentalists.

From this perspective, we can see that Resilient Environmentalists are able to repurpose their capabilities regardless of whether the threat is distant or close. When there are no immediate threats, the business can rely on sensitivity to operations (Weick et al., 2005), situational awareness (Lee et al., 2013) and monitoring (Hollnagel et al., 2006). The broad multi-dimensional resilience of the Resilient Environmentalist enables the organisation to consider a comprehensive meaning of all possible threats for the organisation, whether that be financial, environmental or social through constructive sensemaking (Lengnick-Hall et al., 2011). On the other hand, when a threat is immediate then resilience is faced with responding to an actual disruption and ensuring adaptation and/or bouncing back to "normality" (Boin & van Eeten, 2013; Linnenluecke, 2017). An organisation dealing with potential threats or actual disruptions can become rigid to avoid its operations being knocked off-line for as little time as possible. However, Resilient Environmentalists, because of their multi-dimensional resilience, are more likely to respond with innovation that equally protect the organisation's financial and environmental interests (see e.g., Tuazon et al., 2021).

5.   Future research agenda, recommendations and conclusion

It is clear from this brief foray into OES and OR research that there are overlaps and differences between the two constructs and that it would be short-sighted to look only at an overlapping subset. By gathering information on OR and on OES, we have illustrated four profiles of OESR. We have highlighted some ways in which these quadrants might differ in terms of the OR processes they use, the way in which they perceive OES, and the possible consequences of being in each of the quadrants. Where to next?

From a research perspective, the possibilities are wide open. Although we have generated some ideas around the four profiles in this chapter, we actually know little about why organisations end up with one profile rather than another: Is it to do with sector and industry characteristics? How does business strategy affect the profile? How do organisations move between the profiles and what affects their particular trajectory? and so on.

When we look at the extant literature, it is clear that we have focused a great deal on Resilient Environmentalists as being the "ideal" form and on Survivors as being the "dominant" form. However, research into the downsides of Resilient Environmentalists is necessary to ensure that we have a complete and nuanced view of this important quadrant. A critical perspective on our perception of Survivors should also be taken to identify opportunities to build multi-dimensional resilience into their thinking. Finally, future research needs to examine Self-Sacrificers in particular, to understand how they can be encouraged and supported to become more resilient.

Our definitions of both OR and OES highlighted their meso-level nature and the performance of actors (who have micro-level characteristics and behaviours) using organisational (macro-level) practices and contexts. There is clearly, therefore, a need for a multi-level model that explains the four profiles of OESR and links antecedents and processes to outcomes. The development of such a model is beyond the realm of this chapter, but we encourage future theorists and scholars to take that step.

From a practical perspective, there are also many questions that need answering. Our approach suggests that organisations need to initially determine their overarching goals before making strategic OR and OES decisions; yet empirical research is needed to validate this. Intervention evaluations, action learning, and other participatory research methods will help us to unpack the processes, performances and assumptions that are inherent within each profile.

In summary, therefore, we highlight the importance of considering OR and OES as distinct constructs rather than merely focusing on the overlap between them. With this in mind, we hope that future researchers and practitioners can elucidate each of the quadrants in more detail and put flesh on the skeleton we have laid out here.

References

Augusta, A., Souza, A., Fernandes, M., Alves, R., Macini, N., Cezarino, L. O., & Liboni, L. B. (2017). Resilience for sustainability as an eco-capability. *International Journal of Climate Change Strategies and Management*, 9(5), 1756–8692. https://doi.org/10.1108/IJCCSM-09-2016-0144

Avery, G. C., & Bergsteiner, H. (2011). Sustainable leadership practices for enhancing business resilience and performance. *Strategy & Leadership, 39*(3), 5-15.

Battistella, C., Cicero, L., & Preghenella, N. (n.d.). Sustainable organisational learning in sustainable companies. *The Learning Organization, 28*(1), 15-31. https://doi.org/10.1108/TLO-05-2019-0074

Boin, A., & van Eeten, M. J. G. (2013). The resilient organization. *Public Management Review, 15*(3), 429–445. https://doi.org/10.1080/14719037.2013.769856

Britt, T. W., Shen, W., Sinclair, R. R., Grossman, M. R., & Klieger, D. M. (2016). How much do we really know about employee resilience? *Industrial and Organizational Psychology, 9*(2), 378–404. https://doi.org/10.1017/iop.2015.107

Burgelman, R. A., Floyd, S. W., Laamanen, T., Mantere, S., Vaara, E., & Whittington, R. (2018). Strategy processes and practices: dialogues and intersections. *Strategic Management Journal, 39*(3), 531–558. https://doi.org/10.1002/smj.2741

Carmeli, A., Dothan, A., & Boojihawon, D. K. (2019). Resilience of sustainability-oriented and financially-driven organizations. *Business Strategy and the Environment, 29*(1), 154–169.

Carmeli, A., Friedman, Y., & Tishler, A. (2013). Cultivating a resilient top management team: the importance of relational connections and strategic decision comprehensiveness. *Safety Science, 51*(1), 148–159. https://doi.org/10.1016/j.ssci.2012.06.002

Carvalho, H., & Cruz-Machado, V. (2011). Integrating lean, agile, resilience and green paradigms in supply chain management (LARG_SCM). In P. Li (Ed.), *Supply Chain Management* (pp. 27-48). London: InTech. https://www.intechopen.com/chapters/15530

Dilchert, S., & Ones, D. S. (2021). Environmental sustainability in and of organizations. *Industrial and Organizational Psychology, 5*, 503–511. https://doi.org/10.1111/j.1754-9434.2012.01489.x

Duchek, S. (2020). Organizational resilience: a capability-based conceptualization. *Business Research, 13*(1), 215–246. https://doi.org/10.1007/s40685-019-0085-7

Elkington, J. (1994). Towards the sustainable corporation: win–win–win business strategies for sustainable development. *California Management Review, 36*, 90–100.

Espiner, S., & Becken, S. (2014). Tourist towns on the edge: conceptualising vulnerability and resilience in a protected area tourism system. *Journal of Sustainable Tourism, 22*(4), 646–665.

Espiner, S., Orchiston, C., & Higham, J. (2017). Resilience and sustainability: a complementary relationship? Towards a practical conceptual model for the sustainability-resilience nexus in tourism. *Journal of Sustainable Tourism, 25*(10), 1385–1400. https://doi.org/10.1080/09669582.2017.1281929

Fiksel, J. (2003). Designing resilient, sustainable systems. *Environmental Science & Technology, 37*(23), 5330-5339.

Fisher, D. M., Ragsdale, J. M., & Fisher, E. C. S. (2019). The importance of definitional and temporal issues in the study of resilience. *Applied Psychology, 68*(4), 583–620. https://doi.org/10.1111/apps.12162

Hamel, G., & Välikangas, L. (2003). The quest for resilience. *Harvard Business Review, 81*(9), 52–63.

Hartmann, S., Weiss, M., Newman, A., & Hoegl, M. (2020). Resilience in the workplace: a multilevel review and synthesis. *Applied Psychology, 69*(3), 913–959. https://doi.org/10.1111/apps.12191

Hollnagel, E. (2010). *How resilient is your organisation?* [paper presentation] 31 May to 1 June, Sustainable Transformation: Building a Resilient Organization, Toronto, Canada. https://hal-mines-paristech.archives-ouvertes.fr/hal-00613986/document

Hollnagel, E., Woods, D. D., & Leveson, N. (Eds.) (2006). *Resilience Engineering: Concepts and Precepts*. London: Ashgate Publishing.

Hopkins, D. (2014). The sustainability of climate change adaptation strategies in New Zealand's ski industry: A range of stakeholder perceptions. *Journal of Sustainable Tourism, 22*(1), 107-126.

Hopkins, D. (2015). Applying a comprehensive contextual climate change vulnerability framework to New Zealand's tourism industry. *AMBIO, 44*, 110-120.

Huang, W., Chen, S., & Nguyen, L. T. (2020). Corporate social responsibility and organizational resilience to COVID-19 crisis: An empirical study of Chinese firms. *Sustainability, 12*(21), 8970.

Johnson, N., & Elliott, D. (2011). Using social capital to organise for success? A case study of public–private interface in the UK Highways Agency. *Policy and Society, 30*(2), 101–113. https://doi.org/10.1016/j.polsoc.2011.03.005

Kahn, W. A., Barton, M. A., Fisher, C. M., Heaphy, E. D., Reid, E. M., & Rouse, E. D. (2018). The geography of strain: organizational resilience as a function of intergroup relations. *Academy of Management Review, 43*(3), 509–529. https://doi.org/10.5465/amr.2016.0004

Kollmuss, A., & Agyeman, J. (2002). Mind the gap: why do people act environmentally and what are the barriers to pro-environmental behavior? *Environmental Education Research, 8*(3), 239–260. https://doi.org/10.1080/13504620220145401

Kossek, E. E., & Perrigino, M. B. (2016). Resilience: a review using a grounded integrated occupational approach. *Academy of Management Annals, 10*(1), 729–797. https://doi.org/10.1080/19416520.2016.1159878

Kouamé, S., & Langley, A. (2018). Relating microprocesses to macro-outcomes in qualitative strategy process and practice research. *Strategic Management Journal, 39*(3), 559–581. https://doi.org/10.1002/smj.2726

Kruglanski, A. W., Pierro, A., & Sheveland, A. (2011). How many roads lead to Rome? Equifinality set-size and commitment to goals and means. *European Journal of Social Psychology, 41*, 344–352.

Larsson, M., Milestad, R., Hahn, T., & von Oelreich, J. (2016). The resilience of a sustainability entrepreneur in the Swedish food system. *Sustainability (Switzerland), 8*(6), 550. https://doi.org/10.3390/su8060550

Lee, A. V., Vargo, J., & Seville, E. (2013). Developing a tool to measure and compare organizations' resilience. *Natural Hazards Review, 14*(1), 29–41. https://doi.org/10.1061/(asce)nh.1527-6996.0000075

Lengnick-Hall, C. A., Beck, T. E., & Lengnick-Hall, M. L. (2011). Developing a capacity for organizational resilience through strategic human resource management. *Human Resource Management Review, 21*(3), 243–255. https://doi.org/10.1016/j.hrmr.2010.07.001

Limnios, E. A. M., Mazzarol, T., Ghadouani, A., & Schilizzi, S. G. (2014). The resilience architecture framework: four organizational archetypes. *European Management Journal, 32*(1), 104-116.

Linnenluecke, M. K. (2017). Resilience in business and management research: a review of influential publications and a research agenda. *International Journal of Management Reviews, 19*(1), 4–30. https://doi.org/10.1111/ijmr.12076

Mari, S. I., Lee, Y. H., & Memon, M. S. (2014). Sustainable and resilient supply chain network design under disruption risks. *Sustainability, 6*(10), 6666-6686.

McCarthy, I. P., Collard, M., & Johnson, M. (2017). Adaptive organizational resilience: an evolutionary perspective. *Current Opinion in Environmental Sustainability, 28*, 33-40.

Orlitzky, M., Siegel, D. S., & Waldman, D. A. (2011). Strategic corporate social responsibility and environmental sustainability. *Business and Society, 50*(1), 6–27. https://doi.org/10.1177/0007650310394323

Ortiz-de-Mandojana, N., & Bansal, P. (2016). The long-term benefits of organizational resilience through sustainable business practices. *Strategic Management Journal, 37*(8), 1615–1631. https://doi.org/10.1002/smj.2410

Parsons, D. (2010). Organisational resilience: Parsons argues that a modern organisation's ability to survive and prosper against the flow-on effects of natural disasters will depend on its resilience capacity. *Australian Journal of Emergency Management, 25*(2).

Pieniazek, P. (2017). *Organisational resilience: an investigation into its conceptualisation, outcomes, and boundary conditions* [PhD thesis, University of Leeds]. White Rose Online Repository. https://etheses.whiterose.ac.uk/17745/

Pipe, T. B., Buchda, V. L., Launder, S., Hudak, B., Hulvey, L., Karns, K. E., & Pendergast, D. (2012). Building personal and professional resources of resilience and agility in the healthcare workplace. *Stress and Health, 28*(1), 11–22. https://doi.org/10.1002/smi.1396

Powley, E. H. (2009). Reclaiming resilience and safety: resilience activation in the critical period of crisis. *Human Relations, 62*(9), 1289–1326. https://doi.org/10.1177/0018726709334881

Powley, E. H., Vogus, T. J., Barrett, F. J., Barton, M. A., Dothan, A., Carmeli, A., Sluss, D., & Sutcliffe, K. M. (2017). Making the case for relational resilience. *Academy of Management Proceedings, 2017*(1), 11694. https://doi.org/10.5465/ambpp.2017.11694symposium

Saurin, T. A., & Werle, N. J. B. (2017). A framework for the analysis of slack in socio-technical systems. *Reliability Engineering and System Safety, 167.* https://doi.org/10.1016/j.ress.2017.06.023

Smith, B. W., Dalen, J., Wiggins, K., Tooley, E., Christopher, P., & Bernard, J. (2008). The brief resilience scale: assessing the ability to bounce back. *International Journal of Behavioral Medicine, 15*(3). https://doi.org/10.1080/10705500802222972

Strauss, K., Lepoutre, J., & Wood, G. (2017). Fifty shades of green: how microfoundations of sustainability dynamic capabilities vary across organizational contexts. *Journal of Organizational Behavior, 38*(9), 1338–1355. https://doi.org/10.1002/job.2186

Sutcliffe, K. M., & Vogus, T. J. (2003). Organizing for resilience. In K. S. Cameron, J. E. Dutton, & R. E. Quinn (Eds.), *Positive Organizational Scholarship: Foundations of a New Discipline* (pp. 94–110). San Francisco, CA: Berrett-Koehler.

Tuazon, G. F., Wolfgramm, R., & Whyte, K. P. (2021). Can you drink money? Integrating organizational perspective-taking and organizational resilience in a multi-level systems framework for sustainability leadership. *Journal of Business Ethics, 168*(3), 469–490.

Unsworth, K. L., Davis, M. C., Russell, S. V., & Bretter, C. (2021). Employee green behaviour: How organizations can help the environment. *Current Opinion in Psychology, 42*, 1–6.

Unsworth, K. L., Yeo, G., & Beck, J. (2014). Multiple goals: a review and derivation of general principles. *Journal of Organizational Behavior, 35*(8), 1064–1078. https://doi.org/10.1002/job.1963

van der Vegt, G. S., Essens, P., Wahlström, M., & George, G. (2015). Managing risk and resilience. *Academy of Management Journal, 58*(4), 971–980. https://doi.org/10.5465/amj.2015.4004

Vogus, T. J., & Sutcliffe, K. M. (2007). The impact of safety organizing, trusted leadership, and care pathways on reported medication errors in hospital nursing units. *Medical Care, 45*(10), 997–1002.

Weick, K. E., Sutcliffe, K. M., & Obstfeld, D. (2005). Organizing and the process of sensemaking. *Organization Science, 16*(4), 409–421.

Weick, K. E., Sutcliffe, K. M., Obstfeld, D., Sutton, R. S., & Staw, B. M. (1999). Organizing for high reliability: processes of collective mindfulness. In R. S. Sutton & B. M. Staw (Eds.), *Research in Organizational Behavior* (Vol. 1, pp. 81–123). Stamford, CT: Elsevier Science/JAI Press.

West, B. J., Patera, J. L., & Carsten, M. K. (2009). Team level positivity: investigating positive psychological capacities and team level outcomes. *Journal of Organizational Behavior, 30*(2), 249-267.

Williams, T. A., Gruber, D. A., Sutcliffe, K. M., Shepherd, D. A., & Zhao, E. Y. (2017). Organizational response to adversity: fusing crisis management and resilience research streams. *Academy of Management Annals, 11*(2), 733–769. https://doi.org/10.5465/annals.2015.0134

Windle, G., Bennett, K. M., & Noyes, J. (2011). A methodological review of resilience measurement scales. *Health and Quality of Life Outcomes, 9,* 8. https://doi.org/10.1186/1477-7525-9-8

5 Just transition: the tension between work, employment and climate change

Jo Cutter, Vera Trappmann and Dunja Krause

1. Introduction

The relationship between work, labour and climate change is contentious. Workplaces contribute to large shares of greenhouse gas (GHG) emissions globally. Moving towards reductions in GHG with greater urgency to achieve net-zero by 2050 will require fundamental shifts in socio-political structures, technological and economic systems, organisational forms, cultural values and personal identities, but equally a restructuring of the labour market, generating instability in employment. The transition effects on workers and their responses that in earlier periods of industrial restructuring were seen over decades are now expected to take place within a generation. Resistance to the need for action on climate change has diminished and policy prescriptions are increasingly framed around the jobs and employment co-benefits of investing in a 'green economy' (Stern, 2006) yet pathways to achieve this remain contested (Somerville, 2021), and while the overall effects of decarbonisation on employment are anticipated to be small (see below), the structural change at sectoral and regional levels affect workers and communities unevenly.

Organised labour has lobbied for a 'just transition' to balance the potential tensions created by climate change mitigation strategies, especially in regions where economic deprivation that already exists makes transition more difficult. For organised labour, it is key to combine climate justice with social justice. The International Labour Organization (ILO) states that to minimise the negative effects likely to be felt by already vulnerable groups, any transition needs to be 'managed well' (ILO, 2015, p.4) and respect the four 'key pillars of decent work: social dialogue, social protection, rights at work and employment'. The ILO propose that frameworks for just transitions should anticipate the impacts of sustainability policy on employment and consider adequate social protections and skills programmes alongside social dialogue mechanisms and the rights of workers to organise collectively. Trade unions have developed specific strategies to protect workers in this transition. These range from those reliant upon technological fixes and market solutions, to more autonomous mutual gains strategies. Others advocate for more radical solutions that challenge existing social relations. Within this range of demands, the concept of a just transition has been relatively successful, at least in international policy circles. But understanding what a just transition is, and for whom, has different meanings for stakeholders at different levels (international, national, workplace). In this chapter, we explore the concept of a just transition through the tensions between work, employment and climate change and how transitions towards net-zero (or beyond) affect work and workers. We consider how workers have been involved in transitions to date, drawing on the lessons learned from previous periods of industrial restructuring especially in key carbon-intensive sectors. We explore the specific role of workers and trade unions in the evolution of the 'just transition' narrative. We give examples of how unions try to achieve this in practice and then identify further research directions.

2.   Work, employment and climate change

Work and labour are closely tied to the environment. An estimated 1.2bn of the jobs worldwide (just over one third) such as farming, fishing and forestry rely directly on a healthy environment (ILO, 2018a). Environmental damage stemming from climate change has negative impacts on health, jobs and productivity (ILO, 2019). The increasing frequency and intensity of environment-hazards and extreme weather events, of which heat stress is only one, are already reducing productivity; between 2000 and 2015, an estimated 23 million working-life years were lost annually at the global level as a result of these hazards (ILO, 2018a) and projected temperature increases (even if constrained by climate change measures) will increase these vulnerabilities. On the other hand, industrial work activities lead to environmental damage and climate change. The Intergovernmental Panel on Climate Change (IPCC, 2022) estimates that 35% of GHG emissions are released by the energy sector. The agriculture, forestry and land use sector had seen a decline in GHG emissions but still contribute 22% of (net) global emissions. A further 22%, 15% and 6% of GHGs are contributed by the industry, transport and the building sectors respectively. Apportioning emissions from electricity and heat production to final users (indirect emissions), the shares of industry and the building sectors rise to 34% and 16%. Consequentially, climate change mitigation across many sectors will affect workers.

The 'greening' of the economy implies considerable changes to industrial activity leading to a consequent change in jobs and skills needed. Making estimates of the social and employment effects of climate mitigation policy is difficult as employment outcomes allied to climate change plans depend upon local (country) contexts, industry mix, the low-carbon pathways adopted and implementation infrastructures. Similarly, climate change adaptations are not taking place in isolation and the trajectories employment change are also affected by developments in digitalisation and automation of work and the effects of the global pandemic. Each of these factors is likely to have varied effects for jobs and skill requirements. Several studies suggest that greening the economy will lead to net employment gains. Chateau et al. (2018) suggest an overall increase of 0.2% of employment in the Organisation for Economic Co-operation and Development (OECD) and 0.8% in non-OECD countries between 2011 and 2035. Aggregate job projections, however, mask losses and gains. The ILO project low-carbon strategies would lead to the loss of 6 million jobs by 2030 worldwide alongside the creation of 24 million 'green' jobs in the same time period, a net gain of 18 million jobs (ILO, 2018b). Modelling different low-carbon pathways, New Climate Economy (2018) estimated that 28 million jobs would be lost and 65 million created, a net gain of 37 million (an increase of 0.5–1%) of jobs globally.

Yet, carbon-intensive industries such as mining, fossil fuel and steel production are often localised in regions dependent upon the jobs and incomes they generate, and around which worker and community identities are often forged (McLachlan et al., 2019). For example, in the UK, deep decarbonisation will disproportionately affect the North of England where installations such as coal-fired power stations are concentrated, with the potential loss of relatively high skilled, unionised and well-paid jobs which presents additional threats to regions already facing inequalities (Emden and Murphy, 2018). Thus, the effects of climate change mitigation have differential effects and may exacerbate the vulnerabilities of different groups of workers and their communities.

A further body of research considers the ways in which existing jobs will become green(er) or transformed as new knowledge and skills are required to secure transitions and operate in a low-carbon future, stressing the need for a 'topping up' of skills (Jagger et al., 2013). Using data collated from the O*NET database of standardised and occupation-specific descriptors in the US, Bowen et al. (2018) identified a continuum of jobs that are 'greening'. This ranges from green skills in high demand, to jobs that require reskilling and new jobs emerging. Applying these descriptors to UK labour-market data and projections for low-carbon transitions, Robins et al. (2019) estimate that one in five (6.3 million) UK jobs are either 'transitions aligned' or 'transitions exposed':

3 million jobs (one in ten of the total) face increased demand (aligned) and the remaining 3.3 million jobs require different skills or potential redundancy or redeployment (exposed). Still, workers are ready for this greening, they embrace decarbonisation and expect even more meaningful work from it (Cutter et al., 2021).

The growth of 'green' jobs, however, does not necessarily imply that these will be decent jobs (ILO, 2011; OECD and CEDEFOP, 2014). Existing work-forces in energy and electricity generation are often relatively highly skilled and well unionised jobs but can face greater barriers to accessing new skills and training (OECD, 2012). Concerns of the potential impact on the quality of 'post-displacement' jobs reflect how workers in previous periods of rapid industrial restructuring experienced the lowering of qualities such as pay, pension benefits, job status and autonomy (Botta, 2019). In sectors such as agriculture, mining, transport and manufacturing, low-qualified and older workers tend to be over-represented and it is unlikely that employment and skills systems will respond quickly enough to provide these workers with new skills given the uncertainties involved in low-carbon transitions and weak-nesses in skill formation systems (Jagger et al., 2013). Overall, there may be a risk of the transition to a decarbonised economy to increase the exploitation of labour and put downward pressure on wages (Kenis and Lievens, 2016).

3.   Organised labour response to climate change and net-zero climate policy

Just transition as a concept has been mainstreamed in climate policy and is mentioned in the preface to the 2015 Paris Agreement (UNFCCC, 2015). This was re-affirmed at the United Nations Climate Change Conferences (COPs) in COP24 Katowice[1] and COP26 Glasgow.[2] Just transition has been given prominence at international, national, regional and sectoral policy levels. For example, in the UK's Committee on Climate Change (UK CCC) report on pro-gress to net-zero in 2019 (Ekins, 2019), the need for a just transition is outlined. The report presents just transition primarily as a technical policy objective and as a tool to build public support for net-zero climate policy. This ignores the broader historical and political context in which workers in carbon-intensive and polluting industries have fought to improve working environments and campaign for more effective regulation to address environmental harm (see Hampton, 2015; Morena et al., 2020). The tendency to downplay the agency of workers in policy narratives presents workers and communities as acquiescent victims of economic and industrial change (Hampton, 2015; Barca, 2019).

Yet the inculcation of just transition into wider climate strategy can be clearly attributed to the labour movement. Early iterations of the United Nations' climate change strategy, such as the Kyoto Protocol, were criticised as being 'employment blind' (Rosemberg, 2010, p.128). The mid-2000s saw the emergence of just transition demands and guiding principles at the global level through a series of Trades Union Congress (TUC) federation (ITUC, ETUC, TUC), global union and ILO research and policy reports (ILO, 2010; ILO, 2015; ITUC, 2017; UNECE and ILO, 2020). These built on earlier labour environmental activism and policy work within country-specific union movements (ILO and SustainLabour, 2010; Räthzel and Uzzell, 2012; Stevis and Felli, 2014). This activity undoubtedly built on wider claims for climate justice, but also developed from workplace and community struggles against poor environmental working and living conditions. For instance, key union officials representing workers in the Oil, Chemical and Atomic Workers Union (OCAW) engaged in an 'environmental strike' in US Shell refineries. Working with community campaigners on a related consumer boycott in 1973 they made demands for improved health and safety against toxic hazards and for environmental legislation (Gordon, 1998). OCAW officials also argued for a 'just transition fund' to support workers with the adjustments needed when environmental regulations would lead to the 'sunsetting' of certain industries (Hampton, 2015).

Just transition as a union strategy can be placed within a broader framework of union positions on climate change (Felli, 2014) where labour presents an autonomous perspective, particularly critiquing the 'win–win' argument of so called 'deliberative' positions. Felli argues that through deliberative strategies, unions, non-governmental organisations (NGOs) and employers express shared interests in market-based environmentalism typified within the Kyoto Protocol. This weaker approach to ecological modernisation is contrasted to a more collaborative position within which calls for a just transition are nested (Snell and Fairbrother, 2011; Räthzel and Uzzell, 2012; Hampton, 2015; Brand and Niedermoser, 2019). This approach often critiques excessive growth but advances just transition as a form of state, labour and business collaboration to support necessary change, promoting the interests of workers within existing social relations. Some labour movements also advocate a third, more radical, position that problematises profit, growth and consumption and advocates for new forms of social relations (Clarke and Lipsig-Mummé, 2020). These include campaigns for public ownership of energy as a route to rapid and just decarbonisation (NUMSA, 2012; Sweeny and Treat, 2018) and re-purposing industrial production for the public good, an idea particularly resonant during COVID-19 (Wainwright, 2020).

Notwithstanding these contrasting union positions, just transition has emerged as a policy prescription, adopted more widely as a vehicle through which to address social issues within wider climate policy. As such, just transition as a concept has also become more diffuse and there are differing perspectives on for whom 'justice' should be served. For example, UNFCCC processes underpinning the Paris Agreement centre on national action plans (via the nationally determined contributions or NDC process[3]), but justice also needs to be seen from an international perspective. At the global level, workers are impacted by policy that addresses climate change caused by historic carbon-based practices in the Global North. This raises the issue of so-called common but differentiated responsibility (CBDR) which recognises that while all states have a responsibility to act on climate change, that given different states of development, not all have the same responsibilities. But there continue to be heated debates on this theme. For example, during COP26 in Glasgow some commentators saw the China/India opposition to stronger wording on coal phase-out and the ending fossil fuel subsidies as a reaction to overall attempts of the Global North (especially the US) to move away from CBDR language and equity considerations (ODI, 2021). Other dimensions of justice centre on gender and racial dimensions where, for example, existing union narratives can be criticised for being encased in narrowing down 'justice' around (male) unionised industrial jobs, marginalising the interests of precarious (often young, female, black, migrant) workers (Barca, 2019). There is also a risk of just transition being appropriated by the business community; some reports (for example, B Team, 2018) draw on the language of social dialogue and decent work, but centre on managing risk (and regulation), technology (including skills needs and worker cooperation), markets and reputational damage. This places just transition within the well-established (profit-driven) business narrative, framed as a 'win–win' or, as Moussu (2020, p.69) terms, a 'painless and limited evolution' of business priorities for addressing social issues through including considerations of work and workers within corporate sustainability. There is a risk of taking the significant challenges of climate change and converting them into 'mundane business as usual' (Wright and Nyberg, 2017, p.1633). These varied aspects of equity and justice, and for whom, illustrate the competing interests of political actors engaged in just transition debates at different scales.

4.   Learning from previous transitions

Organised labour's claim for just transition is also based on the experience of previous deep economic transformations that include continuous restructuring as a means of adaptation to globalisation (Strangleman, 2017), or crisis-led restructuring in the aftermath of the financial crisis in 2008 (Doerflinger and Pulignano, 2018). Each often brings displacement or job losses for workers. While the management rationale for restructuring is adaptation to globalisation, improving efficiency and productivity to increase competitive advantage (Kochan et al., 1986; McKinley and Lin, 2012), for workers at the shop-floor level, change and the threat of redundancy has become a normalised experience (McLachlan et al., 2019). Organised labour has developed a variety of strategies to protect their constituencies, from the confrontational to the cooperative (Lévesque and Murray, 2010). Kelly (1996) has argued that militant trade union policies are more likely to secure union survival and recovery compared with moderate policies, identifying more militant 'confrontational job protection' union strategies versus 'collaborative job transition' approaches. The variation in union responses depends on different institutional settings and national employment relations (Frege and Kelly, 2004; Meardi et al., 2009), but also on institutional differences in the labour market (Pulignano, 2011). Existing labour-market systems and regulation that empower displaced workers with training and retraining programmes that support transition see unions seeking agreement with management about redundancies, accepting that redundancies are inevitable but should be organised in a fair way. In contexts that lack labour-market instruments to empower workers, the 'protect jobs' response is a defensive approach towards restructuring, where local actors try to avoid redundancies altogether, looking for difficult internal compromises relying mostly on wage flexibility. In some countries, the effects of restructuring have been particularly bad for workers. In the UK, unions have failed to prevent plant closures and redundancies in the steel industry particularly (Mackenzie et al., 2006; Bacon and Blyton, 2007). In this context, rapid decarbonisation now presents an opportunity to (re-)negotiate justice for workers but the goodwill in unions needs to be rebuilt to avoid a repetition of failures seen during earlier periods of industrial restructuring.

Local union responses can also vary within the same institutional context (Pulignano and Stewart, 2012). Ideology also plays a huge role here (Bacon and Blyton, 2004; Hyman, 2005) as different union orientations of trade unions shape and enable worker capacity to campaign for and secure a just transition. Instruments to mitigate costs for workers in previous rounds of restructuring were realised through: collective bargaining (Marginson, 2015); company-level collective agreements (Glassner and Keune, 2010) and short time work (Glassner and Keune, 2012); redeployment through transfer agencies supporting transitionary employment and skills development (Trappmann and Stuart, 2004); technical training schemes (see Leisink and Greenwood, 2007); and funding programmes for reskilling (Stuart and Wallis, 2007; Stuart et al.,

2013) or the global adjustment (Stuart et al., 2007). It is therefore necessary to grasp the environment in which union strategies emerge, combining a focus on firm-level (local) context together with an assessment of structural (exogenous) and socio-political (endogenous) conditions that foster the production of such power resources (Pulignano and Stewart, 2012). Tonkin (2000) also calls for more attention to the agency of non-organised labour movements. Avoiding the detrimental impacts of massive and rapid industrial restructuring that is driven by plans to decarbonise will require a social contract that includes new systems of social support and security.[4] This could require radical deviation from the way welfare is currently organised based on labour income. Some of the approaches of organised labour towards this more recent transition that focus on embedding justice within climate change policy are under study but this is a nascent field (see Morena et al., 2020). Two cases from our own research[5] illustrate the ways in which workers in mature industrial economies are seeking to advance climate change initiatives.

5.   Organised labour responses to decarbonisation

5.1   IG Metall: transformation short worker allowance – bridging the transformational process

The general position of the metalworkers' union regarding climate change action can be summarised as technology-driven. The transformation is portrayed as an opportunity for competitive advantage in future technology markets where the union's main task is seen to support their workers in the transition and to oppose the re-location of German industrial production abroad (Industriegewerkschaft Metall, 2019b). One of the main policies aimed at facilitating the socially acceptable transformation of German industry is the so-called 'Transformationskurzarbeitergeld' (TKG) (Industriegewerkschaft Metall, 2019a). The TKG is an adapted version of the 'short time worker allowance' which has been used to deal with falling demand in periods of social, economic or financial crisis. Workers reduce their working hours, whilst the state helps to mitigate income losses through financial support. The TKG extends the use of shorter working hours to the transition of entire industries. Workers reduce their working hours to participate in further training and re-qualification programmes for a period of up to two years. The aim of the proposal is two-fold: reduce the potential loss of employment and equip workers with the skills needed in future industries, thereby fostering a continuation of German industry (Industriegewerkschaft Metall, 2019c). The proposal has been supported by the new government, introducing a so-called 'training pay' that will facilitate just transitions.

5.2   Y&H TUC Low-Carbon Taskforce – connecting regional stakeholders including workers and major employers in high polluting sectors

The Yorkshire and Humber (Y&H) region has one of the highest rates of industrial CO_2 emissions per capita in the UK (ETUC, 2014; BEIS, 2021). The effects of an 'unjust' transition towards decarbonisation are already being felt with the relatively sudden closures of the last deep-coal mine in 2016 and major coal-fired power stations in 2016 and 2018 (Prospect, 2020; Diski et al., 2021), alongside continued uncertain futures for the region's high carbon-intensive steel and cement industries. The TUC has brought together unions and employers in the major polluting sectors alongside environmental scientists, local authorities, green think tanks and environmental groups with the aim of connecting regional and local economic development plans to emerging 'transition plans', notably where local authorities have declared a climate emergency and established local climate commissions. The Low-Carbon Taskforce[6] developed as an attempt to build a new kind of institution in the context of limited space for social dialogue in the UK. By promoting the interests of workers and communities the TUC seeks to (re-)build and the role of unions within regional economic policy making in arenas such as the Yorkshire and Humber Climate Commission. Furthermore, the emergence of the Taskforce develops the capacity of union officers and reps in three ways: to foster understanding of and input to regional climate change initiatives such as carbon capture, utilisation and storage (CCUS), industrial hydrogen and off-shore wind, to negotiate just transition clauses within collective agreements and be more proactive in responding to the needs of regional workers requiring support with redundancy or reskilling (TUC, 2020).

The cases above illustrate union action at different levels – those focused on the workplace and on negotiation with employers, to sectoral initiatives and those engaging with public authorities to integrate worker interests in 'green' policy making. While there has been considerable focus at the international level on embedding the principles of a just transition into climate politics, initiatives at the national, sectoral and workplace levels are largely focused on market logic, stressing the role of dialogue between organised labour, business and government, and emphasising workforce training to support the process of adjustment. The lack of more formalised structures of dialogue, notably in the UK, mean that workplace initiatives are especially vulnerable to changes in operational priorities in periods of economic uncertainty, such as the down-turn resulting from the COVID-19 pandemic.

6.   Outlook and future research

In considering the tensions between work, employment and climate change, this chapter highlights a growing understanding between stakeholders of the need to consider the nature of transitions and the role of labour, particularly in carbon-intensive sectors. This is more concretely embedded into policy at the international level. The evolution of just transition, particularly as a policy objective within transnational institutions, can be seen as evidence of effective international union political strategies seeking to ensure that the interests of workers and their communities are considered in wider climate plans previously considered 'blind' to the work and employment effects of climate strategies. The development of the concept of just transition, building from its origins in workplace struggles, particularly those affecting workers in 'frontline' polluting industries, draws attention to the relationship between labour, work and extraction, and autonomous worker demands for justice in the fight for decent working and living conditions. In scaling up these demands from the workplace through to internationally focused arenas, like UNFCCC and COP, these demands have also become more diffuse. When mediated through the interests of other stakeholders (climate activists demanding more rapid transitions, nations of the Global South demanding recognition of past climate injustices and corporate business shaping international governance mechanisms within a market logic), the tensions that emerge when consider-ing the question of work, employment and justice within the actions on climate change require closer attention.

Far-reaching social and economic change cannot be a top-down process: whilst international governance arrangements provide a framework for action, the scale and pace of change demands action by stakeholders at all levels. As the IPCC states, 'social institutions…promote trust and reciprocity, establish networks, and contribute to the evolution of common rules…(and) collective action is reinforced when social actors understand they can participate in local solutions to a global problem that directly concerns them' (IPCC, 2014, p.255). With respect to the role of labour, varied policy innovations are taking place, alongside the emergence of new institutions that promote social dialogue around the concept of a just transition. What is less clear is the specific role(s) for business in securing a 'just transition' for workers at the level of individual business strategy *and* at regional and sectoral levels and how this is effectively developed in dialogue with workers and their representatives. The past demonstrates that we cannot develop strategies that have implications for how the economy works without the engagement of workers and so key questions for future research are:

- How are jobs, skills and labour processes changing in relation to decarbonisation strategies and sustainable business models? How are workers and unions involved in those changes?
- What kinds of 'just transitions' are taking place within organisations in both carbon-intensive sectors and elsewhere?
- Within organisations, what participation and engagement measures are most successful? What is the role of trade unions within effective employee engagement and participation strategies?
- What kinds of participation and democratic structures are needed to engage workers more directly in just transition plans (at organisational, regional, sectoral, national or international levels)?
- Under what conditions can worker participation be sustained at different levels and be effective in securing real progress to achieve climate goals while promoting decent work?

These questions are best addressed by case studies looking at companies, regions and/or sectors on the nature of transitions taking place, the features of effective programmes and the role of workers in both shaping and evaluating their success. More longitudinal analysis through quantitative and qualitative research is also needed. We should not isolate studies on work and employment from climate change but seek interdisciplinary collaboration wherever possible. Including the perspective on work, labour and workers is a pre-requisite to inform policy making to ensure decarbonising happens in a just way.

Notes

1. https://news.industriall-europe.eu/Solidarity and Just Transition Silesia Declaration.pdf.
2. https://ukcop26.org/supporting-the-conditions-for-a-just-transition -internationally/.
3. https://unfccc.int/process-and-meetings/the-paris-agreement/the-paris -agreement/nationally-determined-contributions-ndcs.
4. See for example the work of UNRISD on the eco-social contract: www.unrisd.org/ ib11.
5. These case studies are based on research by Trappmann and Cutter carried out in 2019/20 comparing trade unions responses towards climate change in the UK and

Germany. This involved a review of literature focused on organised labour and climate action, analysis of union conference motions and policy over a 15 year period and 20 expert interviews with senior trade union officials.
6.   https://www.tuc.org.uk/news/low-carbon-task-force-yorkshire-and-humber.

References

B Team 2018. *Just transition. A business guide.* The Just Transition Centre and the B Team. Brussels.

Bacon, N. and Blyton, P. 2004. Trade union responses to workplace restructuring: Exploring union orientations and actions. *Work, Employment and Society.* **18**(4), pp.749–773.

Bacon, N. and Blyton, P. 2007. Conflict for mutual gains? *Journal of Management Studies.* **44**(5), pp.814–834.

Barca, S. 2019. Labour and the ecological crisis: The eco-modernist dilemma in western Marxism(s) (1970s–2000s). *Geoforum.* **98**, pp.226–235.

BEIS 2021. Local authority carbon dioxide emissions estimates 2005–2019. (August). https://geoportal.statistics.gov.uk/datasets/standard-area-measurements-2019-for-administrative-areas-in-the-.

Botta, E. 2019. *A review of Transition Management strategies: Lessons for advancing the green low-carbon transition.* OECD Green Growth Papers, 2019 04. Paris.

Bowen, A., Kuralbayeva, K. and Tipoe, E.L. 2018. Characterising green employment: The impacts of 'greening' on workforce composition. *Energy Economics.* **72**, pp.263–275.

Brand, U. and Niedermoser, M.K. 2019. The role of trade unions in social-ecological transformation: Overcoming the impasse of the current growth model and the imperial mode of living. *Journal of Cleaner Production.* **225**, pp.173–180.

Chateau, J., Bibas, R. and Lanzi, E. 2018. *Impacts of green growth policies on labour markets and wage income distribution: a general equilibrium application to climate and energy policies.* OECD iLibrary.

Clarke, L. and Lipsig-Mummé, C. 2020. Future conditional: From just transition to radical transformation? *European Journal of Industrial Relations.* **26**(4), pp.351–366.

Cutter, J., Trappmann, V., Balderson, U., Sudmant, A. and Norman, H. 2021. *Worker perceptions of climate change and the green transition.* CERIC, University of Leeds. Leeds.

Diski, R., Chapman, A. and Kumar, C. 2021. *Powering the Just Transition.* New Economics Foundation. London.

Doerflinger, N. and Pulignano, V. 2018. Crisis-related collective bargaining and its effects on different contractual groups of workers in German and Belgian workplaces. *Economic and Industrial Democracy.* **39**(1), pp.131–150.

Ekins, P. 2019. *Report to the Committee on Climate Change of the Advisory Group on costs and benefits of net zero.* https://www.theccc.org.uk/wp-content/uploads/2019/05/Advisory-Group-on-Costs-and-Benefits-of-Net-Zero.pdf.

Emden, J. and Murphy, L. 2018. *Risk or reward? Securing a just transition in the north of England – Interim report.* Institute for Public Policy Research. London.

ETUC 2014. Industrial regions and climate policies: Towards a just transition? European Trade Union Confederation. Brussels.

Felli, R. 2014. An alternative socio-ecological strategy? International trade unions' engagement with climate change. *Review of International Political Economy.* **21**(2), pp.372–398.

Frege, C. and Kelly, J. (eds.) 2004. *Varieties of unionism: Strategies for union revitalization in a globalizing economy.* Oxford University Press. Oxford.

Glassner, V. and Keune, M. 2010. *Negotiating the crisis? Collective bargaining in Europe during the economic downturn.* Working Paper No. 10. International Labour Office. Geneva.

Glassner, V. and Keune, M. 2012. The crisis and social policy: The role of collective agreements. *International Labour Review.* **151**(4), pp.351–375.

Gordon, R. 1998. 'Shell no!': OCAW and the labor–environmental alliance. *Environmental History.* **3**(4), pp.460–487.

Hampton, P. 2015. *Workers and trade unions for climate solidarity: Tackling climate change in a neoliberal world.* Routledge. London.

Hyman, R. 2005. Trade unions and the politics of the European social model. *Economic and Industrial Democracy.* **26**(1), pp.9–40.

ILO 2010. Climate change and labour: The need for a 'just transition'. **2**(2), pp.1–264.

ILO 2011. *Skills for green jobs: A global view synthesis report.*

ILO 2015. *Guidelines for a just transition towards environmentally sustainable economies and societies for all.*

ILO 2018a. *The employment impact of climate change adaptation: Input document for the G20 Climate Sustainability Working Group.*

ILO 2018b. *World employment social outlook 2018: Greening with jobs.*

ILO 2019. *Working on a warmer planet: The impact of heat stress on labour productivity and decent work.*

ILO and SustainLabour 2010. *The impact of climate change on employment: Management of transitions through social dialogue.*

Industriegewerkschaft Metall 2019a. Sicher durch die Transformation. *metallzeitung 2019*, p.10.

Industriegewerkschaft Metall 2019b. *Das Transformations-kurzarbeitergeld.*

Industriegewerkschaft Metall 2019c. Wir gestalten den Fairen Wandel. *Transformation*, pp.4–9.

IPCC 2014. *Climate change 2014: Mitigation of climate change. Contribution of Working Group III to the Fifth Assessment Report of the Intergovernmental Panel on Climate Change.* Cambridge University Press. Cambridge, UK and New York.

IPCC 2015. *AR5 Synthesis Report: Climate Change 2014 — IPCC Contribution of Working Groups I, II and III to the Fifth Assessment Report of the Intergovernmental Panel on Climate Change.* Geneva. IPCC.

IPCC 2022. *Summary for Policymakers: Climate Change 2022 — Mitigation of Climate Change. Contribution of Working Group III to the Sixth Assessment Report of the Intergovernmental Panel on Climate Change* [P.R. Shukla, J. Skea, R. Slade, A. Al Khourdajie, R. van Diemen, D. McCollum, M. Pathak, S. Some, P. Vyas, R. Fradera, M. Belkacemi, A. Hasija, G. Lisboa, S. Luz, J. Malley, (eds.)]. Cambridge University Press. Cambridge, UK and New York.

ITUC 2017. *Just Transition – Where are we now and what's next? A guide to national policies and international climate governance.* ITUC Climate Justice Frontline Briefing 2017.

Jagger, N., Foxon, T. and Gouldson, A. 2013. Skills constraints and the low carbon transition. *Climate Policy.* **13**(1), pp.43–57.

Kelly, J. 1996. Union militancy and social partnership. *In*: P. Ackers, C. Smith and P. Smith (eds.) *The new workplace and trade unionism*. Routledge. London, pp.41–76.

Kenis, A. and Lievens, M. 2016. Greening the economy or economizing the green project? When environmental concerns are turned into a means to save the market. *Review of Radical Political Economics*. **48**(2), pp.217–234.

Kochan, T.A., McKersie, R.B. and Chalykoff, J. 1986. The effects of corporate strategy and workplace innovations on union representation. *ILR Review*. **39**(4), pp.487–501.

Leisink, P. and Greenwood, I. 2007. Company-level strategies for raising basic skills: A comparison of Corus Netherlands and UK. *European Journal of Industrial Relations*. **13**(3), pp.341–360.

Lévesque, C. and Murray, G. 2010. Understanding union power: Resources and capabilities for renewing union capacity. *Transfer: European Review of Labour and Research*. **16**(3), pp.333–350.

Mackenzie, R., Stuart, M., Forde, C., Greenwood, I., Gardiner, J. and Perrett, R. 2006. 'All that is solid?': Class, identity and the maintenance of a collective orientation amongst redundant steelworkers. *Sociology*. **40**(5), pp.833–852.

Marginson, P. 2015. Coordinated bargaining in Europe: From incremental corrosion to frontal assault? *European Journal of Industrial Relations*. **21**(2), pp.97–114.

McKinley, W. and Lin, J.C.J. 2012. Executive perceptions: Probing the institutionalization of organizational downsizing. *In*: C.L. Cooper, A. Pandey and J. Quick (eds.) *Downsizing: Is less still more?* Cambridge University Press. Cambridge, pp.228–257.

McLachlan, C.J., MacKenzie, R. and Greenwood, I. 2019. The role of the steelworker occupational community in the internalization of industrial restructuring: The 'layering up' of collective proximal and distal experiences. *Sociology*. **53**(5), pp.916–930.

Meardi, G., Marginson, P., Fichter, M., Frybes, M., Stanojevićánd, S. and Tóth, A. 2009. The complexity of relocation and the diversity of trade union responses: Efficiency-oriented foreign direct investment in Central Europe. *European Journal of Industrial Relations*. **15**(1), pp.27–47.

Morena, E., Krause, D. and Stevis, D. (eds.) 2020. *Just transitions: Social justice in the shift towards a low-carbon world*. Pluto Press. London.

Moussu, N. 2020. Business in just transition: The never-ending story of corporate sustainability. *In*: E. Morena, D. Krause and D. Stevis (eds.) *Just transitions: Social justice in a shift towards a low-carbon world*. Pluto Press. London, pp.56–75.

New Climate Economy 2018. *Unlocking the inclusive growth story of the 21st century: Accelerating climate action in urgent times*. New Climate Economy. Washington DC.

NUMSA 2012. *Motivations for a socially-owned renewable energy sector*. https://numsa .org.za/2012/10/motivations-for-a-socially-owned-renewable-energy-sector-2012 -10-15/.

ODI 2021. Our thoughts on COP26 – reflections on the Glasgow Climate Pact. https:// odi.org/en/insights/our-thoughts-on-cop26-rolling-insight/.

OECD 2012. *OECD green growth studies*. OECD. Paris.

OECD and CEDEFOP 2014. *OECD green growth studies: Greener skills and jobs*. OECD. Paris.

Prospect 2020. *A just transition plan for the UK power sector*. Discussion Paper October 2020. Prospect. London.

Pulignano, V. 2011. Bringing labour markets 'back in': Restructuring international businesses in Europe. *Economic and Industrial Democracy*. **32**(4), pp.655–677.

Pulignano, V. and Stewart, P. 2012. The management of change: Local union responses to company-level restructuring in France and Ireland – a study between and within countries. *Transfer*. **18**(4), pp.411–427.

Räthzel, N. and Uzzell, D. 2012. *Trade unions in the green economy: Working for the environment.* Routledge. London.

Robins, N., Gouldson, A., Irwin, W., Sudmant, A. and Rydge, J. 2019. *Financing inclusive climate action in the UK: An investor roadmap for the just transition.* Grantham Research Institute on Climate Change and the Environment, London School of Economics and Political Science. London.

Rosemberg, A. 2010. Building a just transition: The linkages between climate change and employment. *International Journal of Labour Research.* **2**(2), pp.125-161.

Snell, D. and Fairbrother, P. 2011. Toward a theory of union environmental politics: Unions and climate action in Australia. *Labor Studies Journal.* **36**(1), pp.83-103.

Somerville, P. 2021. The continuing failure of UK climate change mitigation policy. *Critical Social Policy.* **41**(4), pp.628-650.

Stern, N. 2006. *The economics of climate change: The Stern Review.* Cambridge University Press. Cambridge.

Stevis, D. and Felli, R. 2014. Global labour unions and just transition to a green economy. *International Environmental Agreements: Politics, Law and Economics.* **15**(1), pp.29-43.

Strangleman, T. 2017. Deindustrialisation and the historical sociological imagination: Making sense of work and industrial change. *Sociology.* **51**(2), pp.466-482.

Stuart, M., Cutter, J., Cook, H. and Winterton, J. 2013. Who stands to gain from union-led learning in Britain? Evidence from surveys of learners, union officers and employers. *Economic and Industrial Democracy.* **34**(2), pp.227-246.

Stuart, M., Forde, C., MacKenzie, R. and Wallis, E. 2007. An impact study on relocation, restructuring and the viability of the European Globalisation Adjustment Fund: The impact on employment, working conditions and regional development. European Parliament, Brussels. https://www.europarl.europa.eu/RegData/etudes/etudes/join/2006/385647/IPOL-EMPL_ET(2006)385647_EN.pdf.

Stuart, M. and Wallis, E. 2007. Partnership approaches to learning: A seven-country study. *European Journal of Industrial Relations.* **13**(3), pp.301-321.

Sweeny, S. and Treat, J. 2018. *TUED Working Paper #11: Trade unions and just transition - trade unions for energy democracy.* New York. TUED. https://unionsforenergydemocracy.org/resources/tued-working-papers/tued-working-paper-11/

Tonkin, L. 2000. Women of steel: Constructing and contesting new gendered geographies of work in the Australian steel industry. *Antipode.* **32**(2), pp.115-134.

Trappmann, V. and Stuart, M. 2004. *Lifelong learning in the European steel and metal: Results and good practice examples from the LEARNPARTNER project.* Druckerei CONRAD.

TUC 2020. *Voice and place: How to plan fair and successful paths to net zero emissions. Fifth Draft.* TUC. London.

UNECE and ILO 2020. *Jobs in green and healthy transport: Making the green shift.* Report.

UNFCCC 2015. *Adoption of the Paris Agreement.* Geneva. United Nations. https://unfccc.int/resource/docs/2015/cop21/eng/l09r01.pdf.

Wainwright, H. 2020. From airplane wings to ventilator parts – the Lucas Plan. https://lucasplan.org.uk/2020/05/06/from-airplane-wings-to-ventilator-parts/.

Wright, C. and Nyberg, D. 2017. An inconvenient truth: How organizations translate climate change into business as usual. *Academy of Management Journal.* **60**(5), pp.1633-1661.

6 Business models for sustainability: the current state of the literature and future research directions

Suzana Matoh, Katy Roelich and Jonatan Pinkse

1. Introduction to business models for sustainability

The 20th century was marked by economic and social progress with unsustainable use of natural resources. Human activities including production and consumption of food, energy generation, construction and transportation continue to have negative impacts on the planet such as land degradation, loss of biodiversity, groundwater pollution and the climate crisis. Furthermore, there are different social issues associated with unsustainable production and consumption such as dangerous and unfair working conditions and human rights violations. In 2015, United Nations Member States adopted the *2030 Agenda for Sustainable Development* with 17 Sustainable Development Goals (SDGs), an urgent call for action by all countries to ensure a more sustainable future for people and planet (UN General Assembly 2015). Sustainable development considers interrelationships between people, resources, environment and development. SDG12 specifically focuses on ensuring sustainable consumption and production patterns, which requires companies to change their traditional business logic based on the principles of a linear economy and profit-maximisation and find new solutions that enable sustainable consumption and production patterns. Researchers and practitioners consider business models for sustainability as vehicles to deliver the transformative change (Lüdeke-Freund and Dembek 2017). Without a change in the traditional, linear business model, demand for natural resources is likely to continue increasing, therefore, companies across industries need to accelerate the adoption of business models for sustainability to not only create economic value but also social and environmental value.

Business models for sustainability is the central concept in this book chapter, therefore, it is important to clarify how we understand sustainability at the outset. We apply the holistic perspective of sustainability, which includes interactions between environmental, social and economic aspects considering the need to balance short-term gains with the long-term nature of social and environmental issues (Hahn et al. 2015; Lozano 2008). One of the main criticisms of the existing business practices is that companies tend to focus on short-term profits at the expense of long-term security, environmental protection and social equity (Bansal and DesJardine 2014). As a result, short-termism leads to weak sustainability characterised by incremental rather than transformational organisational changes. If business models for sustainability aim to deliver strong sustainability performance, long-term orientation must be considered.

In this chapter, we provide a comparative analysis of business models for sustainability across three carbon-intensive industries: construction and building, energy, and automotive. We examine the main drivers and barriers to business model innovation for sustainability, the types of business models for sustainability that exist in the selected industries (e.g., one dominant business model or a spectrum of business models) and future research opportunities. The chapter is structured as follows. Section 2 introduces the analytical framework that we developed to help us conduct our analysis of business models for sustainability. It includes drivers and barriers that influence business model innovation for sustainability, resulting in different types of business models for sustainability and the forms of sustainable value they create. Section 3 provides a comparative analysis of business models for sustainability across three carbon-intensive industries and proposes future research directions in respective industries. We selected relevant studies for our comparative analysis based on a more subjective literature review and our research expertise. Our comparative analysis provides a perspective on the phenomenon of business models for sustainability. By applying the analytical framework to three different industries, we show how business models for sustainability have worked or have not worked. Section 4 discusses the main findings from the critical analysis and reflects on the prevailing research methods and their suitability to address new research questions in the field of business models for sustainability. Section 5 concludes this chapter with final remarks about business models for sustainability across the industries.

This chapter can serve as complementary reading to Chapter 8 by Alice Owen and Paul Francisco who developed a four-steps approach for sector-specific research on business and sustainability. The two chapters together offer a new perspective on how to pursue research on business models for sustainability within specific industries.

2.   Analytical framework

In this section, we develop our analytical framework that will help us undertake the comparative analysis of the current literature on business models for sustainability in Section 3. First, we focus on the concept of value and its role in business models for sustainability. Then, we define business model innovation for sustainability and describe different types of business models for sustainability based on classifications developed by Bocken et al. (2014). Finally, we explore different internal and external factors that influence business model innovation for sustainability. The analytical framework is the link between the definition of business models for sustainability that we use (centred around the concept of value) and the literature review.

At the heart of a business model is the concept of value. Traditionally, this was expressed in terms of value proposition (products and/or services offered by the firm), value creation and delivery (activities/processes and resources needed to create and deliver a value proposition), and value capture (profit formula) (Osterwalder and Pigneur 2010). With the emergence of business models for sustainability, the concept of value is perceived in a more holistic way. Sustainable value includes economic value (e.g., profit, return on investment and long-term viability), environmental value (e.g., renewable resources, low emissions and biodiversity), and social value (e.g., equality and diversity, well-being and secure livelihood) (Evans et al. 2017). Furthermore, business models for sustainability allow for new governance forms such as cooperatives and social enterprises, which transcend traditional for-profit business models (Schaltegger et al. 2015). Besides sustainable value, business models for sustainability require a system of value flows among different stakeholders, a value network with a new purpose and collaboration between stakeholders for mutual value creation (Evans et al. 2017).

New business models for sustainability are the outcome of business model innovation (see Figure 6.1), that is, deliberate changes to how companies create value (Lüdeke-Freund et al. 2019). Lüdeke-Freund et al. (2019, p. 105) define business model innovation for sustainability as an improvement of a 'company's ability to create, maintain or regenerate natural, social and economic capital beyond its organisational boundaries by changing the value proposition for its customers and all other stakeholders and/or the way how the value is created, delivered and captured.'

There are different types of business models for sustainability (e.g., circular business models and product-service systems) that have the potential to create

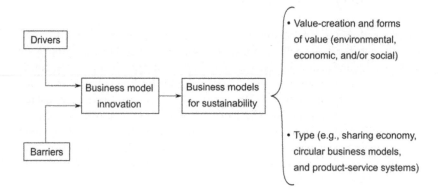

Figure 6.1   Analytical framework

and deliver multiple forms of value (economic, environmental and/or social). Bocken et al.'s (2014) taxonomy of archetypes of business models for sustainability categorised business models based on the dominant type of innovation (technological, social and organisational). There are eight archetypes of business models for sustainability:

(1) Maximise material and energy efficiency (e.g., lean manufacturing and additive manufacturing);
(2) Create value from waste (e.g., circular economy, closed loop, and industrial symbiosis);
(3) Substitute with renewables and natural processes (e.g., move from non-renewable to renewable energy sources);
(4) Deliver functionality rather than ownership (e.g., product-service systems);
(5) Adopt a stewardship role (e.g., biodiversity protection and ethical trade);
(6) Encourage sufficiency (e.g., product longevity and slow fashion);
(7) Repurpose for society/environment (e.g., hybrid businesses); and
(8) Develop scale up solutions (e.g., collaborative approaches).

For the purpose of our analysis, we will consider the taxonomy of archetypes of business models for sustainability developed by Bocken et al. (2014) (see Figure 6.2).

Building on Bocken et al.'s (2014) work, Lüdeke-Freund et al. (2018) classified 45 sustainable business models patterns into 11 groups based on their potential to contribute to value creation along social, environmental and economic

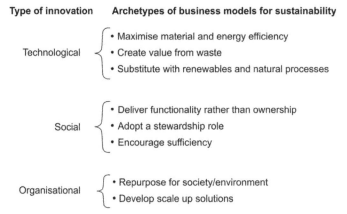

Type of innovation   Archetypes of business models for sustainability

Technological
- Maximise material and energy efficiency
- Create value from waste
- Substitute with renewables and natural processes

Social
- Deliver functionality rather than ownership
- Adopt a stewardship role
- Encourage sufficiency

Organisational
- Repurpose for society/environment
- Develop scale up solutions

Figure 6.2   Business models for sustainability based on Bocken et al.'s (2014) taxonomy of archetypes

dimensions of sustainability. The sustainable business models patterns' groups are:

- Pricing and Revenue Patterns primarily focus on price setting and revenue generation, for example, Freemium and Subscription business models.
- Financing Patterns address the financing aspects within sustainable business models, for example, Crowdfunding, Microfinance and Social business model: no dividends.
- Eco-design Patterns integrate ecological aspects into key activities and value proposition, for example, Hybrid business models and Maximise material productivity and energy efficiency.
- Closing-the-Loop Patterns focus on integrating the concept of circular economy across different business model elements such as key activities, partnerships and customer channels (e.g., Remanufacturing, Repair, Reuse, and Take back management).
- Supply Chain Patterns focus on changing downstream and/or upstream components in a sustainable business model (e.g., Inclusive sourcing, Produce on demand, and Micro distribution and retail).
- Giving Patterns help donate products/services (e.g., Buy one, give one).
- Access Provision Patterns provide products/services for neglected target groups (for example, E-transaction platforms and Experience-based customer credit).
- Social Mission Patterns aim to integrate social target groups in need, for example, Expertise broker and Social business model: empowerment.

- Service and Performance Patterns emphasise the functional and service value of products, for example, Pay for success and Use-oriented services.
- Cooperative Patterns focus on integrating a diverse group of stakeholders as co-owners and co-managers, for example, Cooperative ownership.
- Community Platform Patterns substitute resource/product ownership with community-based access to resources and products, for example, Sharing business.

Different external and internal factors influence business model innovation for sustainability. For example, political and legal factors (e.g., public policies and changes in regulation) and technological development (e.g., technologies that influence sourcing and supply chain transformation) can shape opportunities for new business models (Iles and Martin 2013; Morioka et al. 2017; Sorescu et al. 2011). Furthermore, collaboration with a wide range of stakeholder groups (e.g., engagement with societal actors such as non-governmental organisations, consumers and affected communities) can help companies to design new business models (Carayannis et al. 2015; Iles and Martin 2013). There are also internal factors at the organisational level (e.g., sustainable leadership, vision and continuous innovation) as well as the individual level (e.g., motivations, values and beliefs) that drive business model innovation for sustainability (Høgevold and Svensson 2012; Long et al. 2018; Rauter et al. 2017).

Businesses often encounter barriers when they try to innovate business models. For example, a lack of government support and other external factors such as economic crises can have a negative effect on business model innovation for sustainability (Long et al. 2018). Other important barriers are incumbency and path dependency (Bohnsack et al. 2014). Guldmann and Huulgaard (2020) found that companies are facing barriers at all socio-technical levels, especially the organisational level (e.g., narrow focus of existing sustainability strategies, difficulty attaining management buy-in, and lack of resources and competencies in-house) when trying to develop new circular business models.

To sum up, there are different internal and external factors that can enable or hinder business model innovation for sustainability. The outcomes of the business model innovation process are new types of business models that have the potential to create and deliver economic, social and/or environmental value. In the next section, we apply the analytical framework to help us understand the drivers and barriers for business model innovation for sustainability in the construction and building industry, energy industry and automotive industry. We also analyse the prevailing types of business models for sustainability across the three industries and identify knowledge gaps.

3.   Analysis of business models for sustainability across industrial sectors

In this section, we analyse the current literature on business models for sustainability in the construction and building industry, the energy industry and the automotive industry. This analysis is important because the three industries are carbon-intensive, contributing to environmental and social issues. To ensure sustainable consumption and production patterns in these industries, companies need to accelerate the adoption of business models for sustainability to create social and environmental value and deliver the transformative change.

3.1   Business models for sustainability in the construction and building industry

While the construction and building industry significantly contributes to social and economic benefits (e.g., job creation, productivity and business growth), it is also associated with significant negative impacts, especially GHG emissions. The construction and building industry accounts for 39% of global GHG emissions (International Energy Agency 2017). Efforts to decarbonise the construction and building industry have been made in some leading markets such as the UK, the Netherlands, Sweden and Norway (Construction Climate Challenge 2019). However, more needs to be done to achieve global climate targets. Sustainable practices such as material efficiency, using existing buildings better, switching to low-emission materials, using low-carbon cement, recycling building materials and components, and using low-emission construction machinery (Arup et al. 2019) can be deployed through new business models to accelerate decarbonisation of this industry.

The value chain of construction projects is complex. It involves different stages (design, production, conversion of raw materials into manufactured products and construction itself) with distinct processes and a range of stakeholders (contractors, installers, architects and producers of equipment). The value chain complexity and project-based nature of relationships have influenced a highly fragmented industry structure (Carbon Pricing Leadership Coalition [CPLC] and International Finance Corporation [IFC] 2018) in which traditional business models based on cost reduction rather than performance prevail (Abuzeinab et al. 2017). However, there are construction companies that are championing sustainability initiatives and changing their business models (CPLC and IFC 2018).

There are different drivers and barriers to business model innovation in the construction and building industry. Some of the most critical external and internal drivers are green policy interventions and legislation (e.g., government grants and incentives, mandatory energy efficiency), sustainable technologies and industry structure (e.g., manufacturers and suppliers that provide green materials and products), increasing market demand for sustainable buildings, brand value and public image, entrepreneurship and organisational learning (Zhao et al. 2017). In contrast, key barriers to business model innovation relate to: government constraints such as unadvised policymakers imposing immature regulations; financial constraints such as funding and investment; sector constraints including inherited problems and traditional models; company constraints in relation to focus on short-term profits and lack of vision; and lack of demand for sustainable solutions (Abuzeinab et al. 2017). Furthermore, the process of implementing new business models is challenging, which further constrains business model innovation (Andersson and Lessing 2019; Hossain et al. 2020).

In relation to new business model types in the construction and building industry, product-service systems (Andersson and Lessing 2019; Tseng et al. 2019), sharing economy (Li et al. 2019), and especially circular business models have gained the attention of scholars in recent years. For example, Ünal et al. (2019) explored how value can be created through sustainable managerial practices considering a range of contextual (internal and external) factors. Heesbeen and Prieto (2020) provide an overview of circular business model archetypes (smart input, smart output, stewardship, adaptable building and never-ending building), which can be used to inform design brief and as a roadmap for innovation. Nußholz et al.'s (2019) comparative case study showed that the critical business model innovations for circular materials consider development of partner networks for secondary materials, recovery technologies and capabilities, and identifying customer segments that value lower environmental impact and life cycle costs.

While circular business models are gaining momentum in academic and policy circles, there are doubts about their successful implementation in practice. Therefore, future research should focus on identifying and analysing existing circular business models to better understand success factors and to provide examples of best practice that could be adopted by other companies in the industry. Furthermore, research and practice would benefit from more in-depth case studies focusing on the sectors within the construction and building industry that have the highest potential for GHG emissions reduction such as civil engineering. There are also questions about the decision-making logic that underpins business model innovation, the role of stakeholders that are involved in the complex value chain and the level of sustainability performance achieved due to business model innovation (e.g., weak/strong sustainability and the level of material circularity). Longitudinal studies that focus on the process of business model innovation could help address these questions.

3.2   Business models for sustainability in the energy industry

In many countries, the value chain in the energy sector has emerged as a result of privatisation and liberalisation of a formerly public organisation (Pearson and Watson 2011; Thomas 2006). The necessity to transmit energy through networks means that the flows of energy are separate from the flows of value. The generation, transmission and supply of energy products (gas, electricity and heat) are often undertaken by separate entities (Willis et al. 2019). This affects the flow of value, which goes from generators, who sell to suppliers (often via a wholesale market) then from suppliers to customers (Hall and Roelich 2016). Energy itself moves physically through transmission and distribution networks that are operated by the public sector or by regulated monopolies. The energy sector is highly regulated, particularly the relationship between the energy supplier and the customer (Meyer et al. 2021; Mitchell 2008). It is this history and value chain structure that has shaped the current dominant business model in the energy sector of payment for units of energy and the value proposition to customers is very limited (Hall and Roelich 2016).

There are several drivers for change in this business model. Decarbonisation of the energy sector must happen rapidly to avoid precipitating a climate crisis. Decarbonisation requires the replacement of large, centralised power stations with more distributed renewable technologies and the management of energy demand. This in turn requires new technologies to balance energy generation and demand and the closer participation of customers as prosumers – both producing and consuming energy. The current business model is constraining decarbonisation so alternatives are essential (Hall and Roelich 2016). Fortunately, new technologies are facilitating the creation of new forms of value that can form the basis of more sustainable business models.

These strong drivers for business model innovation are balanced by several barriers, which have the potential to stifle innovation. The strong regulation of markets and focus on economic value can limit the capture of social and environmental value (Hiteva and Foxon 2021). Regulation for consumer protection can also limit some business models directly, including service contracts and peer-to-peer trading (Brown et al. 2019). The uncertainty around future regulation, for example changes in Feed in Tariffs, can act as a barrier itself (Herbes et al. 2017), as can uncertainty about consumer engagement (Shomali and Pinkse 2016). The perception of risk, in such a high-stakes and complex sector, can act as a barrier in a sector dominated by large, risk-averse incumbents (Herbes et al. 2017). The high upfront investment required to procure new technology, or meet regulatory requirements, can further exacerbate this perception of risk.

In the face of these competing drivers and barriers, business model innovation in the energy sector is slow (Hall et al. 2020). However, there are some examples of business models for sustainability. These cannot necessarily be categorised as being a particular type of business model but do create new forms of value. These new forms of value accrue to the energy *system* as well as to customers and the capture of this system value presents a particular challenge in the energy sector (Hiteva and Foxon 2021). New value to the customer can be in the form of increasing comfort associated with energy performance certificates or increasing control and engagement, through prosumer arrangements (Brown et al. 2019). Value to the energy system can include grid stabilisation and flexibility, which overcomes the 'death spiral' associated with an increase in decentralisation and intermittency (Specht and Madlener 2019).

The urgent need for decarbonisation is disrupting the traditional, throughput-based business model in the energy industry. The combination of new technology and increasing customer engagement creates new possibilities for business models that create value for the energy system and for customers. However, technology is still at the early stages of development and deployment and further research is needed to test and track the capability of businesses and customers to engage with these technologies. There is no guarantee that increasing engagement of some customers will lead to a wholesale adoption of new business models, in fact, new research shows that only a small proportion of customers would adopt new business models in the current context (Hall et al. 2021). Therefore, further research is needed to examine how to reduce barriers to the adoption of new business models.

3.3   Business models for sustainability in the automotive industry

For many industrialised countries, the automotive industry is of great economic importance because it creates many jobs and significantly contributes to countries' GDP. However, the production and use of cars, too, contribute to myriad adverse environmental impacts (Orsato and Wells 2007). The automotive industry's value chain used to be stable and companies use identical business models to manufacture and deliver cars to customers. Revenue streams are based on selling or leasing cars, car finance, maintenance, spare parts, and second-hand trade (Athanasopoulou et al. 2019). The technological system of car manufacturing is built on three core attributes: all-steel body, internal combustion engine (ICE), and multi-purpose vehicle (Orsato and Wells 2007). Car companies have served the same customer needs – one car for multiple purposes – using similar product and production technology. Due to societal pressure to address the industry's environmental and climate impact, the industry's business model is undergoing fundamental changes. Car companies are moving away from the ICE and, to a lesser extent, the all-steel body.

Government-enforced technological change and the emergence of carsharing platforms are some of the main drivers for current business model change in the automotive industry (Athanasopoulou et al. 2019; Wesseling et al. 2015). In the EU, for example, the only way for companies to comply with stringent CO_2 standards is to sell electric vehicles (EVs) in mainstream markets. Replacing the ICE with an electrical engine affects business models because EVs have restricted functionality due to current battery technology's limitations (Bohnsack and Pinkse 2017; Wesseling et al. 2020). To address range anxiety and a lack of charging infrastructure, companies have started offering specific-purpose vehicles; EVs were initially marketed as affordable urban vehicles or luxury vehicles with a larger battery (Bohnsack et al. 2014). The rise of EVs has also reinforced the industry's servitisation trend; value capture increasingly relies on a range of services in addition to product sales (Kley et al. 2011). Due to the restricted functionality and new opportunities from digital innovation, the market for EVs is experimenting with a 'mobility-as-a-service' business model (Athanasopoulou et al. 2019; Firnkorn and Müller 2012; Wesseling et al. 2020). This business model is based on the archetype of delivering functionality rather than ownership (Bocken et al. 2014) and has received a major impetus from the popularity of carsharing (e.g., car2go and Drivenow) and ride-hailing (e.g., Uber and DiDi) (Athanasopoulou et al. 2019; Zhang et al. 2020). While existing technologies such as EVs and digital platforms are already changing business models, emerging technologies such as autonomous driving and internet-of things (IoT) have the potential to augment servitisation (Athanasopoulou et al. 2019). An increased use of artificial intelligence, machine learning, voice control, sensors, and vehicle-to-everything (VtoX) communication enable personalised services as well as shared mobility where connected cars could start replacing public transport (Athanasopoulou et al. 2019).

Clearly, the dominant business model in the automotive industry is changing, but there are several barriers as well, holding back a fast and radical transition. As mentioned, the automotive industry's technological system has strong path dependencies; car companies tend to only make incremental changes to their business model (Bohnsack et al. 2014). The required systemic innovation is complex due to strong lock-ins: a strong network of suppliers and dealers, a fossil-fuel-based infrastructure, and customer loyalty (Pinkse et al. 2014; Wesseling et al. 2020). Moreover, servitisation might fall victim to the 'service paradox' (Gebauer et al. 2005): increasingly relying on services requires companies to make high initial investments for which short-term returns are uncertain. Even if there is a trend from ownership to usage, mainstream customers remain sceptical about new business models. And, while governments across the globe have been pushing for the electrification for the automotive industry, policy support has been sketchy and inconsistent, leading many companies to adopt a wait-and-see approach (Bohnsack et al. 2015).

After decades of relative stability, the automotive industry's business model is currently being disrupted. Future research will have to track, though, how fast and radical the business model change will be. Even though Tesla has proven to be a formidable challenger, major car companies are currently regaining market share in the EV market. Once they have a stronghold in this new market, they might be able to better leverage their existing assets, thus reducing the radicality of the business model change. There are many questions, too, about the real impact of emerging technologies such as autonomous and connected cars on the industry. There tends to be much optimism and hype around new technologies. Still, the automotive industry has proven to be quite capable in weathering the storm and accommodate new technologies without fundamentally changing their business model.

Table 6.1 summarises the main findings from the analysis of business models for sustainability across the construction and building industry, energy industry and automotive industry. It shows what are the main factors that drive the use of business models for sustainability and how business models for sustainability are currently being exploited in the three industries. It also shows what are the main differences in relation to the business models for sustainability across the industries as opposed to the main similarities (e.g., short-termism and path dependencies).

4.   Discussion

The construction and building, energy and automotive industries are generating positive economic and social impacts on society, while also contributing to environmental degradation. These industries are expected to continue expanding in the future; therefore, it is important to understand what business models for sustainability currently exist within industries and what factors

Table 6.1   The findings from the analysis of business models for sustainability across industrial sectors

Industry	Factors driving the use of business models for sustainability	Exploitation of business models for sustainability	Differences (value chain and prevailing types of business models for sustainability)
Construction and building	Green policy interventions and legislation Sustainable technologies Increased market demand	Slow due to immature regulations, lack of funding/investment, sectoral constraints, lack of vision and lack of demand for sustainable solutions	The value chain is very complex A spectrum of different business models for sustainability
Energy	Decarbonisation Renewable technologies	Slow due to high regulation, uncertainty around future regulation and perception of risk	The value chain is fragmented No particular type of business models for sustainability New forms of value to the system difficult to capture
Automotive	Government-enforced technological change Emergence of carsharing platforms	Slow due to technological system's path dependency, strong lock-ins, customers' scepticism, and inconsistent/inadequate policy support	The value chain is becoming unstable Business models for sustainability based on servitisation (e.g., functionality rather than ownership)

influence business model innovation. Understanding the current situation can help inform future research, practice and policymaking to support more transformative business model innovations for sustainable consumption and production patterns, that is, to achieve SDG12.

Our analysis showed that technological changes and developments are one of the main drivers for business model innovation across the three industries. For example, sustainable materials in construction and building are enabling circular business models. In the energy industry, renewable technologies for energy generation, management and distribution, and participation of prosumers are driving the creation of business model alternatives. In the automotive industry, digital innovations and emerging technologies (e.g., IoT, autonomous driving and machine learning) are enabling personalised services and shared mobility. While technology is an important driver of business model innovation, the pace of business model innovation is slow and the level of radicality is low. Furthermore, financial constraints such as high upfront costs and funding as well as consumers' perceptions (e.g., scepticism about new business models, perceptions of risks and lack of demand) are also hindering business model innovation across the three industries.

In all three industries there are examples of business models for sustainability that create new forms of value for a range of stakeholders. While business models for sustainability in the energy industry cannot necessarily be categorised as being a particular type of business model, there is a trend in the automotive industry towards servitisation and sharing business models. Similarly, a trend can be observed in construction and building towards circular business models.

Based on our analysis, we suggest that future research should focus on investigating the transformative power of new technologies (e.g., substantial organisational change), how companies can speed up the uptake of business models for sustainability, and what the real outcomes are of new business models (economic, social and environmental benefits). The construction and building industry and the energy industry both have complex value chains that include a range of stakeholders. Therefore, future research should consider the role of different stakeholder groups in the business model innovation process and the outcomes of business model innovation for each stakeholder group. Action research with industrial partners, working on specific problems and finding solutions that can be applied across industries would be particularly beneficial. Current research uses mostly qualitative methods such as single case studies or comparative case studies with two cases, and interviews; there are also some examples of mixed methods research. To answer pressing research questions, multiple case studies including stakeholders' workshops could be utilised. Furthermore, in-depth interviews and longitudinal studies are needed to track changes and impacts of business model innovations over time.

5.   Conclusion

This chapter analysed business models for sustainability across three carbon-intensive industries and proposed future research directions. Decarbonisation is one of the main driving factors of business model innovation to ensure sustainable development. However, business model innovation tends to be slow and stifled by barriers at micro, meso and macro levels. It is evident that business models are changing, but we need to better understand the real outcomes of these changes and their contribution to sustainable development. While businesses across industries have a key role to play in driving the SDGs, governments and consumers need to work together with the industry to ensure a sustainable future for all.

References

Abuzeinab, A., M. Arif and M. A. Qadri (2017), 'Barriers to MNEs green business models in the UK construction sector: An ISM analysis', *Journal of Cleaner Production*, **160**, 27–37.

Andersson, N. and J. Lessing (2019), 'Product service systems in construction supply chains', *Periodica Polytechnica Architecture*, **50** (2), 132–8.

Arup, University of Leeds and C40 (2019), *Building and Infrastructure Consumption Emissions*. https:// www .arup .com/ perspectives/ publications/ research/ section/ buildings-and-infrastructure-consumption-emissions

Athanasopoulou, A., M. de Reuver, S. Nikou and H. Bouwman (2019), 'What technology enabled services impact business models in the automotive industry? An exploratory study', *Futures*, **109** (March), 73–83.

Bansal, P. and M. R. DesJardine (2014), 'Business sustainability: It is about time', *Strategic Organization*, **12** (1), 70–8.

Bocken, N. M. P., S. W. Short, P. Rana and S. Evans (2014), 'A literature and practice review to develop sustainable business model archetypes', *Journal of Cleaner Production*, **65**, 42–56.

Bohnsack, R., A. Kolk and J. Pinkse (2015), 'Catching recurring waves: Low-emission vehicles, international policy developments and firm innovation strategies', *Technological Forecasting and Social Change*, **98**, 71–87.

Bohnsack, R. and J. Pinkse (2017), 'Value propositions for disruptive technologies: Reconfiguration tactics in the case of electric vehicles', *California Management Review*, **59** (4), 79–96.

Bohnsack, R., J. Pinkse and A. Kolk (2014), 'Business models for sustainable technologies: Exploring business model evolution in the case of electric vehicles', *Research Policy*, **43** (2), 284–300.

Brown, D., S. Hall and M. E. Davis (2019), 'Prosumers in the post subsidy era: An exploration of new prosumer business models in the UK', *Energy Policy*, **135**, 110984.

Carayannis, E. G., S. Sindakis and C. Walter (2015), 'Business model innovation as lever of organizational sustainability', *Journal of Technology Transfer*, **40** (1), 85–104.

Carbon Pricing Leadership Coalition and International Finance Corporation (CPLC and IFC) (2018), *Construction Industry Value Chain: How Companies are Using Carbon Pricing to Address Climate Risk and Find New Opportunities*, Washington, DC: International Finance Corporation, accessed 2 November 2020 at https:// www .ifc .org/ wps/ wcm/ connect/ topics_ext_content/ ifc_external_corporate_site/ climate+business/resources/construction-industry-value-chain.

Construction Climate Challenge (2019), *Decarbonizing the Construction Industry: Can Europe Lead in Low Carbon Buildings and Infrastructure?*, accessed 10 October 2020 at https:// construc tionclimat echallenge .com/ events/ decarbonizing -the -construction -industry-can-europe-lead-in-low-carbon-buildings-and-infrastructure/.

Evans, S., D. Vladimirova, M. Holgado, K. Van Fossen, M. Yang, E. A. Silva and C. Y. Barlow (2017), 'Business model innovation for sustainability: Towards a unified perspective for creation of sustainable business models', *Business Strategy and the Environment*, **26** (5), 597–608.

Firnkorn, J. and M. Müller (2012), 'Selling mobility instead of cars: New business strategies of automakers and the impact on private vehicle holding', *Business Strategy and the Environment*, **21** (4), 264–80.

Gebauer, H., E. Fleisch and T. Friedli (2005), 'Overcoming the service paradox in manufacturing companies', *European Management Journal*, **23** (1), 14–26.

Guldmann, E. and R. D. Huulgaard (2020), 'Barriers to circular business model innovation: A multiple-case study', *Journal of Cleaner Production*, **243**, 118160.

Hahn, T., J. Pinkse, L. Preuss and F. Figge (2015), 'Tensions in corporate sustainability: Towards an integrative framework', *Journal of Business Ethics*, **127** (2), 297–316.

Hall, S., J. Anable, J. Hardy, M. Workman, C. Mazur and Y. Matthews (2021), 'Matching consumer segments to innovative utility business models', *Nature Energy*, **6**, 349–36.

Hall, S., C. Mazur, J. Hardy, M. Workman and M. Powell (2020), 'Prioritising business model innovation: What needs to change in the United Kingdom energy system to grow low carbon entrepreneurship?', *Energy Research and Social Science*, **60**, accessed at https://doi.org/10.1016/j.erss.2019.101317.

Hall, S. and K. Roelich (2016), 'Business model innovation in electricity supply markets: The role of complex value in the United Kingdom', *Energy Policy*, **92**, accessed at https://doi.org/10.1016/j.enpol.2016.02.019.

Heesbeen, C. and A. Prieto (2020), 'Archetypical CBMs in construction and a translation to industrialized manufacture', *Sustainability (Switzerland)*, **12** (4), accessed at https://doi.org/10.3390/su12041572.

Herbes, C., V. Brummer, J. Rognli, S. Blazejewski and N. Gericke (2017), 'Responding to policy change: New business models for renewable energy cooperatives – barriers perceived by cooperatives' members', *Energy Policy*, **109**, 82–95.

Hiteva, R. and T. J. Foxon (2021), 'Beware the value gap: Creating value for users and for the system through innovation in digital energy services business models', *Technological Forecasting and Social Change*, **166**, 120525.

Høgevold, N. M. and G. Svensson (2012), 'A business sustainability model: A European case study', *Journal of Business and Industrial Marketing*, **27** (2), 142–51.

Hossain, M. U., S. T. Ng, P. Antwi-Afari and B. Amor (2020), 'Circular economy and the construction industry: Existing trends, challenges and prospective framework for sustainable construction', *Renewable and Sustainable Energy Reviews*, **130** (May), 109948.

Iles, A. and A. N. Martin (2013), 'Expanding bioplastics production: Sustainable business innovation in the chemical industry', *Journal of Cleaner Production*, **45**, 38–49.

International Energy Agency (IEA) (2017), 'Towards a zero-emission, efficient, and resilient buildings and construction sector'. https://www.worldgbc.org/news-media/global-status-report-2017

Kley, F., C. Lerch and D. Dallinger (2011), 'New business models for electric cars: A holistic approach', *Energy Policy*, **39** (6), 3392–403.

Li, Y., R. Ding, L. Cui, Z. Lei and J. Mou (2019), 'The impact of sharing economy practices on sustainability performance in the Chinese construction industry', *Resources, Conservation and Recycling*, **150** (December 2018), 104409.

Long, T. B., A. Looijen and V. Blok (2018), 'Critical success factors for the transition to business models for sustainability in the food and beverage industry in the Netherlands', *Journal of Cleaner Production*, **175**, 82–95.

Lozano, R. (2008), 'Envisioning sustainability three-dimensionally', *Journal of Cleaner Production*, **16** (17), 1838–46.

Lüdeke-Freund, F., S. Carroux, A. Joyce, L. Massa and H. Breuer (2018), 'The sustainable business model pattern taxonomy: 45 patterns to support sustainability-oriented business model innovation', *Sustainable Production and Consumption*, **15**, 145–62.

Lüdeke-Freund, F. and K. Dembek (2017), 'Sustainable business model research and practice: Emerging field or passing fancy?', *Journal of Cleaner Production*, **168**, 1668–78.

Lüdeke-Freund, F., S. Schaltegger and K. Dembek (2019), 'Strategies and drivers of sustainable business model innovation', in F. Boons and A. McMeekin (eds), *Handbook of Sustainable Innovation*, Cheltenham, UK and Northampton, MA, USA: Edward Elgar Publishing Limited, pp. 101–23.

Meyer, D., J. Fauser and D. Hertweck (2021), 'Business model transformation in the german energy sector: Key barriers and drivers of a smart and sustainable transformation process in practice', *ISPRS Annals of the Photogrammetry, Remote Sensing and Spatial Information Sciences*, VIII-4/W1-2021, 73–80.

Mitchell, C. (2008), *The Political Economy of Sustainable Energy*, Basingstoke: Palgrave Macmillan.

Morioka, S. N., I. Bolis, S. Evans and M. M. Carvalho (2017), 'Transforming sustainability challenges into competitive advantage: Multiple case studies kaleidoscope converging into sustainable business models', *Journal of Cleaner Production*, **167**, 723–38.

Nußholz, J. L. K., F. Nygaard Rasmussen and L. Milios (2019), 'Circular building materials: Carbon saving potential and the role of business model innovation and public policy', *Resources, Conservation and Recycling*, **141** (March 2018), 308–16.

Orsato, R. J. and P. Wells (2007), 'U-turn: The rise and demise of the automobile industry', *Journal of Cleaner Production*, **15** (11–12), 994–1006.

Osterwalder, A. and Y. Pigneur (2010), *Business Model Generation: A Handbook for Visionaries, Game Changers, and Challengers*, Hoboken, NJ: John Wiley & Sons.

Pearson, P. and J. Watson (2011), *UK Energy Policy 1980–2010: A History and Lessons Learnt. A Review to Mark 30 Years of the Parliamentary Group for Energy Studies*, London: The Parliamentary Group for Energy Studies.

Pinkse, J., R. Bohnsack and A. Kolk (2014), 'The role of public and private protection in disruptive innovation: The automotive industry and the emergence of low-emission vehicles', *Journal of Product Innovation Management*, **31** (1), 43–60.

Rauter, R., J. Jonker and R. J. Baumgartner (2017), 'Going one's own way: Drivers in developing business models for sustainability', *Journal of Cleaner Production*, **140**, 144–54.

Schaltegger, S., E. G. Hansen and F. Lüdeke-Freund (2015), 'Business models for sustainability: Origins, present research, and future avenues', *Organization & Environment*, **29** (1), 3–10.

Shomali, A. and J. Pinkse (2016), 'The consequences of smart grids for the business model of electricity firms', *Journal of Cleaner Production*, **112**, 3830–41.

Sorescu, A., R. T. Frambach, J. Singh, A. Rangaswamy and C. Bridges (2011), 'Innovations in retail business models', *Journal of Retailing*, **87**, S3–16.

Specht, J. M. and R. Madlener (2019), 'Energy supplier 2.0: A conceptual business model for energy suppliers aggregating flexible distributed assets and policy issues raised', *Energy Policy*, **135**, 110911.

Thomas, S. (2006), 'The British model in Britain: Failing slowly', *Energy Policy*, **34** (5), 583–600.

Tseng, M. L., S. Lin, C. C. Chen, L. S. Calahorrano Sarmiento and C. L. Tan (2019), 'A causal sustainable product-service system using hierarchical structure with linguistic preferences in the Ecuadorian construction industry', *Journal of Cleaner Production*, **230**, 477–87.

UN General Assembly (2015), *Transforming Our World: The 2030 Agenda for Sustainable Development*, accessed 12 November 2020 at https://www.refworld.org/docid/57b6e3e44.html.

Ünal, E., A. Urbinati, D. Chiaroni and R. Manzini (2019), 'Value creation in circular business models: The case of a US small medium enterprise in the building sector', *Resources, Conservation and Recycling*, **146** (December 2018), 291–307.

Wesseling, J. H., C. Bidmon and R. Bohnsack (2020), 'Business model design spaces in socio-technical transitions: The case of electric driving in the Netherlands', *Technological Forecasting and Social Change*, **154** (February 2019), 119950.

Wesseling, J. H., E. M. M. I. Niesten, J. Faber and M. P. Hekkert (2015), 'Business strategies of incumbents in the market for electric vehicles: Opportunities and incentives for sustainable innovation', *Business Strategy and the Environment*, **24** (6), 518–31.

Willis, R., C. Mitchell, R. Hoggett, J. Britton, H. Poulter, T. Pownall and R. Lowes (2019), *Getting Energy Governance Right: Lessons from IGov*, Exeter: IGov, University of Exeter.

Zhang, Y., J. Pinkse and A. McMeekin (2020), 'The governance practices of sharing platforms: Unpacking the interplay between social bonds and economic transactions', *Technological Forecasting and Social Change*, **158** (June), 120133.

Zhao, X., T. Chang, B. G. Hwang and X. Deng (2017), 'Critical factors influencing business model innovation for sustainable buildings', *Sustainability*, **10** (1), 1–19.

7 A research agenda for green supply chain management

Chee Yew Wong and Qinghua Zhu

1. Introduction

According to Boston Consulting Group, there are eight supply chains that account for 50% of global annual greenhouse gas (GHG) emissions: food, construction, fashion, fast-moving consumer goods (FMCG), electronics, automotive, professional services and freight (Burchardt et al., 2021). While research in green supply chain management (GSCM) has revealed practices that reduce environmental and societal impacts, there is no real reduction in GHG emissions globally. This chapter suggests advancing GSCM research in several perspectives. First, the Global North relies on low-tier (the second or lower-tier) upstream suppliers from the Global South (e.g., Latin America, Asia, Africa, and Oceania) for production and material extractions without adequate understanding how their demand and requirements affect their abilities to implement green technologies. Thus, we advocate a multi-tier approach to GSCM. Second, we advocate a shift from a focal-firm (e.g., brands or original manufacturers) and supplier-customer view to consider stakeholders such as governments, regulators, non-governmental organisations (NGOs) and local communities as the focal unit of analysis because they play important roles in GSCM, especially driving activities for achieving the sustainable development goals (SDGs). A multi-stakeholder perspective is proposed to enrich the focal-firm view. Third, we suggest a closer multi-level examination of the motivations, behaviours, and perceptions of responsibilities among important stakeholders, focal firms and their suppliers and customers to understand intra-organisational dynamics. Fourth, instead of a "business as usual" mindset, we suggest a circular supply chain management (SCM) by incorporating the circular economy (CE) concept.

2.   State of the literature

Handfield and Nichols (1999: 2) posit that "supply chains encompasses all activities associated with the flow and transformation of goods from raw materials (extraction), through the end user, as well as associated information flows". The term supply chain also denotes "the network of suppliers, distributors and consumers" (Zhu and Sarkis, 2004: 267). Thus, SCM concerns the management of upstream and downstream relationships (Christopher, 2011). Within a firm, SCM are treated as functions, which consist of procurement (purchasing), operations (production) and logistics functions. Environmental management has become an important research topic in the operations and manufacturing management field since the 1990s (Klassen, 1993). GSCM integrates sustainability issues into SCM research that examines ethical or sustainable procurement (Green et al., 1998), cleaner production of manufacturing (Klassen, 1993), and green logistics (Murphy et al., 1995; Wu and Dunn, 1995).

Based on our experience researching GSCM literature, we present here some critiques of the existing GSCM literature. The term GSCM was introduced (Zhu and Sarkis, 2004) to encompass internal environmental management practices, external green supply chain management, eco-design, and investment recovery. GSCM integrates concepts such as "green supply", "green purchasing", "green logistics", "sustainable operations", "environmental supply chain management" or "environmentally sustainable supply chain management".

GSCM research has been criticised for driving limited impacts. It relies on several theoretical perspectives: resource dependence theory, transaction cost economics, population ecology, and the resource-based view of the firm (Carter and Rogers, 2008) that symbolically add sustainability into an economic-driven paradigm. While incorporating triple bottom line (TBL) into sustainable SCM, Carter and Easton (2011) argue that stakeholders, societies and individuals (people) and the natural environment (planet) should be treated as being as important as profit. SCM emphasises economic values by satisfying customers, while environmental and sustainability issues remain secondary. Moreover, some of the above formal theories that focus on generalisation fail to reveal the complexity and dynamics involved in implementing GSCM. Thus, Carter et al. (2020) further emphasised the needs for middle range theory, conceptual theory building and grounded theory to gain refined understanding.

Another problem is that the focus on a focal-firm perspective means the perspectives of other stakeholders are less understood. Seuring and Müller (2008) suggested it is important for focal firms (e.g., brands or original manufacturer) to understand the perspectives of governments, customers, stakeholders, and suppliers from a multi-tier perspective. They also pointed out sustainable development is often reduced to minor environmental improvement by GSCM research; instead, there is a need for more radical innovation in the ways the natural environment is managed. Pagell and Wu (2009) argue the literature often focuses on modifications of existing practices, which not only promotes small improvement, but also hinders businesses from becoming truly sustainable in the long run. They highlight efforts to reconceptualise the supply chains that consider NGOs, competitors, trade groups and govern-ments so that radical changes in product and supply chain design can be made to meet multiple stakeholder interests. Moreover, Pagell and Shevchenko (2014: 44) argue "previous research has focused on the synergistic and familiar while overlooking trade-offs and radical innovation" and future research should avoid an emphasis on harm reduction (instead of harm elimination), limited stakeholder view (driven by a focus on profit), familiar practices, empirical quantitative methods (instead of using rich qualitative data), and measurement of impacts. The main research question is not "does it pay?", but "how to be sustainable?" (Pagell and Shevchenko, 2014: 49). Following these critiques, we propose a future research agenda below.

3.   Agenda for future research

3.1   Towards a multi-tier approach

The relationship between a supplier and a customer – dyads – has been the fundamental unit of analysis in SCM (Choi and Wu, 2009). Since the focus on dyads does not capture the complexities of a network, it is suggested to regard a supply chain as a multi-tier system, considering there are many tiers of upstream suppliers and downstream customers (Mena et al., 2013). In 2014, Tachizawa and Wong adapted the idea of a multi-tier supply chain to the sustainability context. This idea was inspired by the increasing concerns that suppliers engaging in unsustainable behaviours can become a "chain liability" for a buying firm, as downstream customers may protest or boycott the buying firm as a reaction to the news of such undesirable behaviours (Hartmann and Moeller, 2014). Since significant pollution can be produced by second or lower-tier suppliers and NGOs often use such cases to put pressure on buying firms, there is the need for a multi-tier approach to managing the sustainability of a supply chain (Tachizawa and Wong, 2014). Considering buying firms are often "lead firms", Tachizawa and Wong (2014) argue many such firms choose to only impose sustainability requirements to their first-tier suppliers and expect them to cascade the requirements upstream. Some lead firms may directly work with or influence the lower-tier suppliers or work indirectly with third parties such as local governments, NGOs, competitors, auditors, standards institutions, and so on. They argue such multi-tier relationships are contingent on power structure, stakeholder pressure, material criticality, industry dynamism/pollution level, dependency, distance, and knowledge resources.

The idea of a lead firm within a multi-tier supply chain was extended by Jia et al. (2019) based on case studies of multinationals in China. The case studies show all the direct and indirect approaches proposed by Tachizawa and Wong (2014) are used to manage multi-tier supply chains, such as working with waste collection companies, using training companies to train farmers and tier-2 suppliers, and pass on requirements to tier-2 suppliers via tier-1 suppliers. Jia et al. (2019) find that supply chain leadership is a more appropriate construct than power in a multi-tier GSCM context as some buyers have limited power over lower-tier suppliers. They observe the use of transformational leadership on tier-1 suppliers and transactional leadership on middle-tier suppliers, which influence supply chain learning. Learning can take place due to the use of a third party, such that resource orchestration is enabled by multi-tier and stakeholder interactions. This finding adds evidence about the roles of knowledge resources initially pointed out by Tachizawa and Wong (2014). Thus, more research is required to understand how the use of different multi-tier supply chain approaches leads to different learning.

The study of mineral supply chains by Sauer and Seuring (2019) shows the need to recognise a "cascaded" multi-tier supply chain, whereby there is a global mineral market between upstream focal firms and downstream focal firms. This cascaded structure is also applied to many agri-food global supply chains, and it may prevent direct interactions between the two focal firms, and such a problem leads to the use of third parties such as industry alliances instead. Consequently, theories that are useful for explaining the dyadic relationships between a buyer and a supplier, for example transaction cost economy or agency theory, will need to be extended to accommodate a multi-party unit of analysis (e.g., including a supplier and its suppliers). The traditional view that a buyer would only interact with its first-tier suppliers is not accurate because a buyer may interact or trade with its second-tier suppliers and multiple suppliers for the same input (Mena et al., 2013). Wilhelm et al. (2016: 42) highlighted that first-tier suppliers will act as double agents, first as an agent who needs to satisfy the lead firm's sustainability requirements (i.e., the primary agency role) and second as an agent who implements these requirements in their suppliers' operations (i.e., the secondary agency role). Their case studies show that lead firms need to incentivise each agency role of first- and second-tier suppliers separately, and to reduce information asymmetries (Wilhelm et al., 2016). Such multiple agencies and roles are still less understood.

3.2   Towards a multi-stakeholder approach

The need to consider external stakeholders was highlighted when environmental management was initially incorporated into the field. Klassen (1993: 82) argued "manufacturing must expand its traditional external focus on customers and suppliers to include third-party stakeholders such as government agencies and the public". Stakeholder integration was identified as an important capability for a proactive environmental strategy (Sharma and Vredenburg, 1998). However, the question of how multiple stakeholder concerns can be integrated into supply chain decisions remains less studied. There is a tendency to over-simplify concerns of various external stakeholders into aggregated concepts such as stakeholder pressures (e.g., Kim and Lee, 2012; Sarkis et al., 2010). Even breaking them down into customers, regulators and owners (e.g., Pålsson and Kovacs, 2014) does not add new insights. A similar tendency is observed in the use of institutional theory to understand how institutional pressures (e.g., coercive, normative, and mimetic pressures) drive the adoption of GSCM practices (e.g., Zhu et al., 2013). A deeper appreciation of the perspectives of each salient stakeholder based on a multi-stakeholder perspective is required.

While environmental degradation and climate change affect everyone, many parties have different views on how a supply chain should be made "greener". Governments or regulators are important stakeholders that drive environmental sustainability. Since the 1980s, an increasing number of new environmental protection regulations have been introduced by governments. Based on ecological modernisation theory, Zhu et al. (2011) show that environmental problems due to economic growth can be mitigated by increasing resource efficiency technology innovation-driven governmental policies. Regulations become an important instrument as many governments struggle to meet their commitments to the 2015 Paris Agreement, that is, to cut down carbon dioxide intensity (CO_2 emission per unit of Gross Domestic Product [GDP]) by 2030. According to the Porter hypothesis (PH), well-designed and stringent environmental regulations can stimulate innovations, which in turn increase the productivity of firms or the product value for end users (Porter, 1991). The argument that environmental regulations drive innovation is regarded as a weak version of PH, while it serves as a constraint to profitability effects on business performance as a strong version of PH (Jaffe and Palmer, 1997). Although there is evidence supporting the weak version of PH, it is not enough to significantly drive business performance (van Leeuwen and Mohnen, 2017). In countries treated as "pollution heavens", environmental regulations are less effective in controlling and reducing pollution (Hao et al., 2018). Unless new knowledge is created to confront such a dilemma, many governments would have to choose between economic (job) growth and the use of stringent environmental regulations to ensure environmental sustainability at the expense of slower growth.

One important debate GSCM research can contribute to is the idea of "*decoupling*". The growth in economic activity increases carbon emissions because all economy activities involve energy, production and logistics technologies that produce emissions. Decoupling economic growth and carbon emissions means deep emissions reductions are made possible with little or no effect on economic growth. To understand how decoupling can be realised, there are many research challenges. President Obama referred to the statistics in the 2017 Economic Report of the President (ERP-2107) and claimed that, during the period of his presidency (2008-2015), CO_2 emissions from the energy sector fell by 9.5% while the economy grew by over 10%. There may be a reduction in emissions in a post-industrialised country, but its economy growth is supported by (polluting) production activities from producing countries that export to such consuming countries. Accounting for the total global emissions caused by the consumption of products produced by producing countries requires further research. The question of whether producers who create values should be allowed to capture more values remains contradictory to the interests of multinationals (Kumar, 2019).

The SCM literature has provided knowledge on how to manage factories, international manufacturing networks and outsourcing of production to low-cost countries in the Global South, such as Vietnam, Indonesia, Brazil, Mexico, India, Bangladesh, Kenya. The setting up of factories overseas or sourcing from low-cost countries can also be an exploitation of the "pollution heavens" policies. A country is chosen for building factories partly because the host government uses a "race to the bottom" policy (Porter, 1999), that is, offering low tax and imposing low environmental and labour regulations while labour costs are relatively low. While such a policy promotes global sourcing activities that largely destroy the natural environment, studies about global sourcing often care more about cost efficiency and effectiveness from the perspective of outsourcers or buyers (Stanczyk et al., 2017), rather than its dark sides. Such a narrow perspective must be changed.

Instead, the GSCM literature needs to consider the perspectives of stakeholders, for example, local governments, workers, and their families from the low-cost countries or Global South. One main debate globally is whether it is "morally right" to treat these countries as "pollution heavens" and "let/allow" them to sacrifice environmental quality while achieving economic growth through speedy industrialisation and foreign direct investment without investing in environmental protection. The classical environmental Kuznets Curve (EKC) hypothesis argues that there is an inverted-U-shaped relationship between environmental pollution and average income. At the early stage of the economic development, environmental pollution may increase as the economy grows, but when the GDP per capita is high enough the turning point of environmental pollution would be reached and then environmental pollution can decrease alongside economic growth (Grossman and Krueger, 1995). While there is mixed evidence for the EKC hypothesis (Galeotti et al., 2006) and there are problems accounting for local/global emissions (He and Richard, 2010), the exploitation of such "pollution heavens" and "race to the bottom" policies remain a major reason why global supply chains continue to pollute and some countries remain "stuck in the bottom" (Porter, 1999). Further research is required to reveal such morally wrong policies.

To stop the "race to the bottom", many stakeholders are involved, including those involved in trade agreements, local government, domestic labour unions, multinational and global supply chains (Anner, 2015) and workers (Kumar, 2019). Competition among low-cost countries to attract foreign direct investment (FDI) can lead to the adoption of "pollution heavens" and "race to the bottom" policies. While more recent generations of trade agreements have started to consider social and labour welfare clauses, the enforcement of such clauses requires cooperation between local governments, domestic labour unions and the global supply chains. The evidence that internal institutional pressures are relatively less effective than international institutional pressures in China suggests international regulations and pressures from interstate trade agreements (e.g., WTO) do have an important role to play (Zhu et al., 2012). Furthermore, it is equally important for buying firms, particularly those from the Global North to work with local governments and regulators (Tachizawa and Wong, 2014). Often significant changes only occur after a serious disaster. A relevant example is the collapse of Rana Plaza in 2013. However, the subsequent increases in labour wages, rises of safety requirements in Bangladesh, and continuous practice of price squeeze have pressured multinational companies to shift the production of textile to countries such as Ethiopia (Anner, 2015). It appears impossible to stop the "race to the bottom" while cases of "race from the bottom" remain (Kumar, 2019). GSCM research can contribute by highlighting trade policies that continue to promote the "race to the bottom" and how supply chain companies exploit such policies by interviewing multiple stakeholders from the Global North and Global South.

3.3   Towards a multi-level analysis

An emphasis on focal firms and firm-level analysis means GSCM research fails to understand how decisions are made at an individual level (Carter et al., 2020) and how managers confront trade-offs on a day-to-day basis (Pagell and Wu, 2009). Few studies examine motivations, behaviours, and perceptions of responsibilities among individual supply chain employees. Gattiker et al. (2014) show certain organisational climate and individual values drive supply chain employees to champion sustainable supply chain initiatives. They demonstrate the value of using intra-organisational influence theory (Kipnis et al., 1980; Yukl and Falbe, 1990) to understand how the use of different influencing tactics help gain commitment from employees to contribute to GSCM projects. There are hard tactics that rely on authority and assertive behaviours, and soft tactics that include ingratiation and the use of aspirational appeals (Kipnis et al., 1980; Yukl and Falbe, 1990). Gattiker et al. (2014) show that a lower risk climate makes it easier for employees to commit to GSCM projects.

In addition, Cantor et al. (2012) show top management (supervisory) support and environmental training are important factors that drive employees' commitment to environmental behaviours, for example, increasing the frequency of involvement, but these factors are not adequate to drive innovative environmental behaviours and efforts to promote environmental initiatives. There are other factors affecting proactive championing behaviours (Gattiker et al., 2014). This shows it is not easy to achieve a contagious effect. The finding of Cantor et al. (2012) shows reward did not drive commitment. Individuals may perceive their responsibilities in GSCM as a form of autonomous (self-determined) or controlled motivation, which may be shaped by the environment and individuals' preference or attitudes. The case studies of Murphy et al. (2019) show different intervention pathways shape employees' involvement in GSCM activities. Top management can promote GSCM by deploying resources to implement such initiatives, provide training, disseminate information through two-way communication, stimulate employees' individual responsibility, relatedness and efforts to participate in solution finding and engagement with external parties. The question remains whether an employee treats environmental championing as part of his/her job, or it is better to assign sustainability roles to every employee.

Even though Cantor et al. (2013) show it is important to gain environmental commitment of the individual responsible for environmental management practices, not all employees are assigned such a responsibility. This means organisations rely on other forms of influence, such as targets and policies or employees who would voluntarily engage in GSCM. Recent evidence from

Verma et al. (2018) shows some employees with autonomous motivation acquire responsibility and become proactive when they are given the opportunity or assigned the responsibility to drive sustainability initiatives in their functions. All these studies suggest the formation of employee responsibility is simultaneously affected by organisational contexts and individual attitudes and behaviours. To gain a more complete understanding of how GSCM is formed or changed in an organisation, further multi-level studies are required (Carter et al., 2015).

3.4   Challenging "business as usual" mentality

Regarding the multi-tier, multi-stakeholder and/or multi-level analysis that is needed for GSCM studies and implementation, innovative ideas and approaches are put forward below to drive more in-depth studies.

3.4.1   From GSCM to circular SCM

GSCM studies, extending to multi-tier suppliers, mainly focus on forward supply chains while at the same time closed-loop SCM considers both forward and reverse supply chain operations (Batista et al., 2018). From a firm's perspective, GSCM literature can be complemented by the idea of circular economy to achieve environmental and economic performance simultaneously (Liu et al., 2018). Thus, circular supply chain management (CSCM) has been proposed (Elia et al., 2020; Farooque et al., 2019). Like closed-loop SCM, CSCM aims to increase the recovery rate and the level of used products (Atabaki et al., 2020). Besides environmental performance, CSCM focuses on value creation (Mishra et al., 2018). Thus, CSCM extends from the existing reuse, remanufacturing, and recycling in the same supply chain of new products to other supply chains (Kazancoglu et al., 2020).

CSCM needs to redesign supply chains along the multi-tier, but difficulties exist. Based on a literature review and multiple case studies, Bressanelli et al. (2019) identifies twenty-four challenges that may hamper design for CSCM, which can be grouped into seven categories: economic and financial viability, market and competition, product characteristics, standards and regulation, supply chain management, technology, and users' behaviour. Sixteen of these challenges have been well studied in related fields such as closed-loop supply chains or servitisation. However, the remaining eight are new, including the market cannibalisation between circular and new products, the impact of fashion changes, the standards and regulations-related challenges (taxation and policy instruments misalignment, metrics, lack of standards), cultural issues specific to the CE, data privacy and security at the end-of-use, and the willingness to pay for reclaimed products. Besides, technology innovation is needed for CSCM. Industry 4.0 technologies have brought opportunities for CSCM (Singh et al., 2019), but barriers exist (Ozkan-Ozen et al., 2020).

To implement CSCM, new business models are required. De Angelis et al. (2018) argue that circular economy principles should be integrated into GSCM, including a shift from product ownership to leasing, which brings different relationships along multi-tier supply chains. Yang et al. (2018) further put forward product-service business models (Yang et al., 2018). Considering possible value creation in inner circles, circling long and cascading use circles, they introduced three types of product-service CSCM business models (i.e., product-, use- and result-oriented models). Due to the complexity of processes and activities in circular supply chains, risk identification and management are crucial (Ethirajan et al., 2020). To overcome difficulty in implementing CSCM, we need multi-level cooperation and collaboration within one firm and among firms in multiple supply chains.

3.4.2   The role of NGOs and other stakeholders

Due to the increasing risks in low-tier suppliers (Sedex Global and Partners, 2013), researchers have extended GSCM studies to low-tier or multi-tier suppliers, but the dominant studies are about how to manage Tier-2 suppliers through Tier-1 suppliers. GSCM, especially for CSCM and multi-tier GSCM, requires strong collaboration among stakeholders along supply chains (De Angelis et al., 2018). NGOs are one of the key stakeholders in GSCM. In most cases, NGOs disclose information when they find environmental violations. Such violation disclosure will attract other concerned stakeholders, such as consumers and investors. Thus, violation disclosure drives market share loss not only for firms with violations but also customers of the firms. For instance, Lo et al. (2018) find that when two stakeholders (governments and NGOs) publicise environmental violations by Chinese manufacturing suppliers, both the suppliers and their overseas customers suffer a significantly negative stock market reaction.

Disclosure of environmental violations by NGOs can affect the environmental reputation of firms and even their customers. However, understanding the other loss of supply disruption, many multi-national customers have supported NGOs to scrutinise their suppliers' environmental responsibility. For instance, the Institute for Public & Environmental Affairs (IPE), which is famous for actively scrutinising local firms' environmental performance and linking them to gigantic customers, revealed environmental violations of Apple's suppliers through the "Bloody Apple" campaign (https://finance .huanqiu.com/article/9CaKrnJsjIe). However, Apple donated funds to the IPE via its sponsored foundation – Society of Entrepreneurs & Ecology (SEE) Foundation. Besides financial support, many leading customers, such as Apple and Nike, have publicised their list of suppliers, and thus NGOs and the media can check their suppliers' environmental performance (Chen et al., 2019). Kraft et al. (2020) further studied why and how customers should disclose social responsibility information to consumers.

For the conventional approach, customers usually keep supplier information confidential to avoid potential reputation loss due to environmental violations of their suppliers. Now, customers have realised that they need to balance reputation loss and disruption loss. Moreover, more and more customers have tried to help suppliers to be more environmentally responsible by pursuing collaboration with stakeholders such as governments, NGOs and consumers. This suggests the need to better understand information disclosure from an information user perspective.

4.   Final remarks

This chapter suggests several new perspectives to advance GSCM research. The aim is to move GSCM research from focusing on minor improvements towards driving radical innovation. In an era of climate change, business as usual is not an option. A focal-firm centric approach will keep the research focus on profitability rather than the natural environment (planet) and society (people). The focus on dyads as the unit of analysis will not help understand complex issues facing upstream suppliers and other stakeholders and policy makers.

We need to ask new research questions: What drives members of a supply chain to innovate and seriously transform towards a truly sustainable one? What makes supply chains address the tensions between a focal-firm and an environmental/societal centric view? Who are the important stakeholders and policy makers who can help transform a supply chain and how to understand their roles and interactions with the supply chains? How to identify and understand how sustainability requirements are passed on in complex multi-tier supply chains? We hope the above research agenda and questions provide some platforms to transform the research landscape of GSCM from a focal-firm minor environmental improvement mindset to enable multi-tier and multi-stakeholder collaboration to achieve sustainable development goals by multi-level collaboration and cooperation.

Acknowledgements

Qinghua Zhu is supported by the projects of the National Natural Science Foundation of China (72221001, 72192833/72192830, 72088101).

References

Anner, M. 2015. Stopping the race to the bottom: Challenges for workers' right in the supply chains in Asia. *International Policy Analysis, Friedrich Ebert Stiftung*, 1-8.

Atabaki, M.S., Mohammadi, M., Naderi, B. 2020. New robust optimization models for closed-loop supply chain of durable products: Towards a circular economy. *Computer & Industrial Engineering*. 146, 106520.

Batista, L., Bourlakis, M., Smart, P., Maull, R. 2018. In search of a circular supply chain archetype – a content-analysis-based literature review. *Production Planning & Control*. 29, 438–451.

Bressanelli, G., Perona, M., Saccani, N. 2019. Challenges in supply chain redesign for the circular economy: A literature review and a multiple case study. *International Journal of Production Research*. 57, 7395–7422.

Burchardt, J., Fredeau, M., Hadfield, M., Herhold, P., O'Brien, C., Pieper, C., Weise, D. 2021. Supply chains as a game-changer in the fight against climate change, BCG, 26 January 2021, https://www.bcg.com/publications/2021/fighting-climate-change-with-supply-chain-decarbonization

Cantor, D.E., Morrow, P.C., McElroy, J.C., Montabon, F. 2013. The role of individual and organizational factors in promoting firm environmental practices. *International Journal of Physical Distribution & Logistics Management*. 43(5/6), 407-427.

Cantor, D.E., Morrow, P.C., Montabon, F. 2012. Engagement in environmental behaviors among supply chain management employees: An organizational support theoretical perspective. *Journal of Supply Chain Management*. 48(3), 33-51.

Carter, C.R., Easton, P.L. 2011. Sustainable supply chain management: Evolution and future directions. *International Journal of Physical Distribution & Logistics Management*. 41(1), 46-62.

Carter, C.R., Hatton, M.R., Wu, C., Chen, X. 2020. Sustainable supply chain management: Continuing evolution and future directions. *International Journal of Physical Distribution & Logistics Management*. 50(1), 122-146.

Carter, C.R., Meschnig, G., Kaufmann, L. 2015. Moving to the next level: Why our discipline needs more multilevel theorization. *Journal of Supply Chain Management*. 51(4), 94-102.

Carter, C.R., Rogers, D.S. 2008, A framework of sustainable supply chain management: Moving toward new theory. *International Journal of Physical Distribution and Logistics Management*. 38(5), 360-387.

Chen, S., Zhang, Q.Q., Zhou, Y.P. 2019. Impact of supply chain transparency on sustainability under NGO scrutiny. *Production Operation Management*. 28, 3002-3022.

Choi, T.Y., Wu, Z. 2009. Taking the leap from dyads to triads: Buyer–supplier relationships in supply networks. *Journal of Purchasing and Supply Management*. 15(4), 263-266.

Christopher, M. 2011. *Logistics & supply chain management*, 4th ed. FT Prentice Hall.

De Angelis, R., Howard, M., Miemczyk, J. 2018. Supply chain management and the circular economy: Towards the circular supply chain. *Production Planning & Control*. 29, 425-437.

Elia, V., Gnoni, M.G., Tornese, F. 2020. Evaluating the adoption of circular economy practices in industrial supply chains: An empirical analysis. *Journal of Cleaner Production*. 273, 122966, 1-14.

Ethirajan, M., Arasu, M.T., Kandasamy, J., Vimal, K.E.K., Nadeem, S.P., Kumar, A. 2020. Analysing the risks of adopting circular economy initiatives in manufacturing supply chains. *Business Strategy and the Environment*. 30(1), 204-236.

Farooque, M., Zhang, A., Thurer, M., Qu, T., Huisingh, D. 2019. Circular supply chain management: A definition and structured literature review. *Journal of Cleaner Production*. 228, 882-900.

Galeotti, M., Lanza, A., Pauli, F. 2006. Reassessing the environmental Kuznets curve for CO_2 emissions: A robustness exercise. *Ecological Economics*. 57, 152–163.

Gattiker, T.F., Carter, C.R., Huang, X., Tate, W.L. 2014. Managerial commitment to sustainable supply chain management projects. *Journal of Business Logistics*. 35(4), 318-337.

Green, K., Morton, B., New, S. 1998. Green purchasing and supply policies: Do they improve companies environmental performance? *Supply Chain Management*. 3(2), 89-95.

Grossman, G., Krueger, A., 1995. Economic growth and the environment. *The Quarterly Journal of Economics*. 110, 353–377.

Handfield, R.B., Nichols, Jr. E.L. 1999. *Introduction to supply chain management*. Pearson.

Hao, Y., Deng, Y., Lu, Z-N., Chen, H. 2018. Is environmental regulation effective in China? Evidence from city-level panel data. *Journal of Cleaner Production*. 188, 966-976.

Hartmann, J., Moeller, S. 2014. Chain liability in multitier supply chains? Responsibility attributions for unsustainable supplier behavior. *Journal of Operations Management*. 32(5), 281-294.

He, J., Richard, P. 2010. Environmental Kuznets curve for CO_2 in Canada. *Ecological Economics*. 69, 1083-1093.

Jaffe, A., Palmer, K. 1997. Environmental regulation and innovation: A panel data study. *Review of Economics and Statistics*. 79(4), 610–619.

Jia, F., Gong, Y., Brown, S. 2019. Multi-tier sustainable supply chain management: The role of supply chain leadership. *International Journal of Production Economics*. 217, 44-63.

Kazancoglu, I., Sagnak, M., Mangla, S.K., Kazancoglu, Y. 2020. Circular economy and the policy: A framework for improving the corporate environmental management in supply chains. *Business Strategy and the Environment*. 30(1), 590-608.

Kim, S., Lee, S. 2012. Stakeholder pressure and the adoption of environmental logistics practices: Is eco-oriented culture a missing link? *The International Journal of Logistics Management*. 23(2), 238-258.

Kipnis, D., Schmidt, S.M., Wilkinson, A. 1980. Intraorganizational influence tactics: Explorations in getting one's way. *Journal of Applied Psychology*. 65, 440–452.

Klassen, R.D. 1993. The integration of environmental issues into manufacturing: Toward an interactive open-systems model. *Production & Inventory Management Journal*. 34(1), 82–88.

Kraft, T., Valdés, L., Zheng, Y. 2020. Motivating supplier social responsibility under incomplete visibility. *Manufacturing & Service Operations Management*, Published online in Articles in Advance, 11 February 2020.

Kumar, A. 2019. A race from the bottom? Lessons from a workers' struggle at a Bangalore warehouse. *Competition & Change*. 23(4), 346–377.

Liu, J.J., Feng, Y.T., Zhu, Q.H., Sarkis, J. 2018. Green supply chain management and the circular economy: Reviewing theory for advancement of both fields. *International Journal of Physical Distribution & Logistics Management*. 48, 794–817.

Lo, C.K.Y., Tang, C.S., Zhou, Y., Yeung, A.C.L., Fan, D. 2018. Environmental incidents and the market value of firms: An empirical investigation in the Chinese context. *Manufacturing & Service Operations Management*. 20, 422–439.

Mena, C., Humphries, A., Choi, T.Y. 2013. Toward a theory of multi-tier supply chain management. *Journal of Supply Chain Management*. 49(2), 57–77.

Mishra, J.L., Hopkinson, P.G., Tidridge, G. 2018. Value creation from circular economy-led closed loop supply chains: A case study of fast-moving consumer goods. *Production Planning & Control*. 29, 509–521.

Murphy, E., Da Costa, N.G., Wong, C.Y. 2019. Decoding human intervention: Pathways to successful environmental management. *European Management Review*. 17(1), 247–265.

Murphy, P.R., Poist, R.F., Braunschweig, C.D. 1995. Role and relevance of logistics to corporate environmentalism: An empirical assessment. *International Journal of Physical Distribution & Logistics Management*. 25(2), 5–19.

Ozkan-Ozen, Y.D., Kazancoglu, Y., Mangla, S.K. 2020. Synchronized barriers for circular supply chains in Industry 3.5/Industry 4.0 transition for sustainable resource management. *Resource Conservation & Recycling*. 161, 104986.

Pagell, M., Shevchenko, A. 2014. Why research in sustainable supply chain management should have no future. *Journal of Supply Chain Management*. 50(1), 44–55.

Pagell, M., Wu, Z.H. 2009. Building a more complete theory of sustainable supply chain management using case studies of 10 exemplars. *Journal of Supply Chain Management*. 45(2), 37–56.

Pålsson, H., Kovacs, G. 2014. Reducing transportation emission – a reaction to stakeholder pressure or a strategy to increase competitive advantage. *International Journal of Physical Distribution & Logistics Management*. 44(4), 283–304.

Porter, G. 1999. Trade competition and pollution standards: "Race to the bottom" or "stuck at the bottom". *Journal of Environment & Development*. 8(2), 133–151.

Porter, M. 1991. America's Green Strategy. *Scientific American*. 264, 168. doi: 10.1038/scientificamerican0491-168

Sarkis, J., Gonzalez-Torre, P., Adenso-Diaz, B. 2010. Stakeholder pressure and the adoption of environmental practices: The mediating effect of training. *Journal of Operations Management*. 28(2), 163–176.

Sauer, P.C., Seuring, S. 2019. Extending the reach of multi-tier sustainable supply chain management: Insights from mineral supply chains. *International Journal of Production Economics*. 217, 31–43.

Sedex Global and Partners 2013. Going deep: The case for multi-tier transparency. Sedexglobnal.com, https:// www .sedex .com/ wp -content/ uploads/ 2016/ 09/ Sedex -Transparency-Briefing.pdf

Seuring, S., Müller, M. 2008. From a literature review to a conceptual framework for sustainable supply chain management. *Journal of Cleaner Production*. 16, 1699-1710.

Sharma, S., Vredenburg, H. 1998. Proactive corporate environmental strategy and the development of completely valuable organizational capabilities. *Strategic Management Journal*. 19, 729-753.

Singh, S.P., Singh, R.K., Gunasekaran, A., Ghadimi, P. 2019. Supply chain management, Industry 4.0, and the circular economy. *Resource Conservation & Recycling*. 142, 281-282.

Stanczyk, A., Cataldo, Z., Blome, C., Busse, C. 2017. The dark side of global sourcing: A systematic literature review and research agenda. *International Journal of Physical Distribution & Logistics Management*. 47(1), 41-67.

Tachizawa, E.M., Wong, C.Y. 2014. Towards a theory of multi-tier sustainable supply chains: A systematic literature review. *Supply Chain Management: An International Journal*. 19(5/6), 643-663.

van Leeuwen, G., Mohnen, P. 2017. Revisiting the Porter hypothesis: An empirical analysis of Green innovation for the Netherlands. *Economics of Innovation and New Technology*. 26(1-2), 63-67.

Verma, S., Wong, C.Y., Unsworth, K. 2018. Behavioural drivers of employee engagement towards environmental sustainability: A case of UK public sector organisation. In proceedings of 29th Production and Operations Management Society Conference POMS 2018, 4-7 May. Texas.

Wilhelm, M., Blome, C., Bhakoo, V., Paulraj, A. 2016. Sustainability in multi-tier supply chains: Understanding the double agency role of the first-tier supplier. *Journal of Operations Management*. 41, 42-60.

Wu, H., Dunn, S.C. 1995. Environmentally responsible logistics systems. *International Journal of Physical Distribution & Logistics Management*. 25(2), 20-38.

Yang, M.Y., Smart, P., Kumar, M., Jolly, M., Evans, S. 2018. Product-service systems business models for circular supply chains. *Production Planning & Control*. 29, 498-508.

Yukl, G., Falbe, C.M., 1990. Influence tactics and objectives in upward downward and lateral influence attempts. *Journal of Applied Psychology*. 75(2), 132-140.

Zhu, Q., Cordeiro, J., Sarkis, J. 2012. International and domestic pressures and responses of Chinese firms to greening. *Ecological Economics*. 83, 144-153.

Zhu, Q., Geng, Y., Sarkis, J., Lai, K-H. 2011. Evaluating green supply chain management among Chinese manufacturers from the ecological modernization perspective. *Transportation Research Part E*. 47, 808-821.

Zhu, Q.H., Sarkis, J. 2004. Relationships between operational practices and performance among early adopters of green supply chain management practices in Chinese manufacturing enterprises. *Journal of Operations Management*. 22(3), 265-289.

Zhu, Q., Sarkis, J., Lai, K-H. 2013. Institutional-based antecedents and performance outcomes of internal and external green supply chain management practices. *Journal of Purchasing & Supply Management*. 19, 106-117.

8 Researching business and sustainability in specific sectors: the example of the construction industry

Alice Owen and Paul Francisco

1.   Introduction

Businesses in different sectors use different materials, processed in different ways, to provide different products and services. Supply chains may be interwoven between sectors, and the same customers will be served by several business sectors, but to meet different customer needs. Thus, while the same kinds of sustainability issues will arise for many business sectors – for example resource efficiency, environmental impacts, labour conditions, health impacts – the particular characteristics of how that issue is manifest, and what steps can be taken to tackle the issue will vary from sector to sector. The types of challenges facing automotive production, food processing and software development may have common ground, but the practicalities of how to improve sustainability in those sectors will be different. Sector-specific research matters in the field of business and sustainability because it is only sector-specific analysis that will lead directly to sector-specific, effective, action.

This chapter offers a pathway to identifying sector-specific research questions, followed by thoughts on how to answer those research questions. The first point to explore is why such a pathway is necessary; how does it help explore the sustainability issues for a specific sector any more than an approach which involves scanning the existing literature and identifying an interesting research topic? We suggest that using the four steps illustrated in Figure 8.1 gives you the confidence that you will be focusing on research which is valuable, responding to the specific needs of the sector you are studying, and therefore leading to greater confidence in the usefulness and impact of your research.

We use the construction industry as our example to explore sector-specific sustainability research partly because that is the field where we have developed

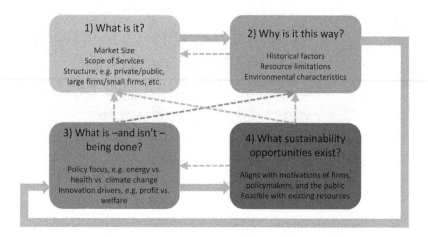

Figure 8.1   Four steps to create a high impact focus for business sustainability research

expertise and partly because the construction industry offers both sustainability challenges and also huge potential for more sustainable outcomes. Construction is both a sector essential to human well-being in providing shelter and a sector which must transform if we are to move closer to sustainable development. The authors' research journeys have taken rather different paths through engineering, design, waste management and policy development to arrive at a similar point: the inescapable importance of professional knowledge, standards, motivations, beliefs and behaviours in delivering more sustainable construction processes and outputs.

Construction covers the design, provision and maintenance of the built environment: infrastructure and buildings, homes and commercial buildings, schools and hospitals. To illustrate our points, we will focus in on one part of this sector, the repair, maintenance and improvement (RMI) of existing residential buildings. RMI is a helpful focus as it is vitally important in delivering one of the main sustainability challenges at this moment in time – the imperative to get to "net zero" carbon emissions. While design and technology is readily available to create new buildings to zero carbon standards, existing buildings will make up around 90% of the 2050 UK building stock (Power, 2010) and so, somehow, these existing buildings need to be transformed to zero carbon; this has to happen through RMI processes. In the US the percentage is expected to be lower, with existing buildings representing about three-fourths of the 2050 US building stock. This is based on a new build rate of approximately 1.5M units per year (U.S. Census Bureau, 2020) and a retirement rate of 0.6% per year (Emrath, 2018).

1.1   Step 1: Characterise the sector

The first step in identifying useful sector-specific research objectives is to develop a deep and multi-faceted understanding of the sector in question.

Our example, the construction sector, is significant for sustainability economically, socially and environmentally. The sector is large both in financial turnover terms with housing-related construction alone accounting for £43bn of economic activity in the UK in 2018 (Office for National Statistics, 2019) and about $600bn in the US in 2019 (National Association of Home Builders, 2020). The construction sector as a whole provides around 6% of the UK's total economic activity (Rhodes, 2019) and about 4% of GDP in the US (Deloitte, 2019). In employment terms, there were 1.3 million construction workers in the UK in 2018 (Office for National Statistics, 2019) and over 8 million in the US (Statista, 2020) meaning that the sector supports a large number of livelihoods.

To understand business and sustainability for a specific sector, the configuration of the organisations in that sector is important. For construction, this means recognising the types of firms, their sizes, their activities, outputs and success measures, together with how dependencies within a network of firms operate, and the underpinning skills and knowledge that enable those firms to function. For example, in the UK, construction firms are either very large or very small, with a sparsely populated middle ground. In the UK, 13% of construction workers work in firms employing more than 1200 people while 38% of construction workers are sole traders or in firms of seven people or less (Office for National Statistics, 2019). Smaller firms typically specialise in one trade, or one type of construction, and operate in networks and teams to deliver projects (Owen, 2015). This reflects the project-based nature of construction work, fundamentally different from many manufacturing processes (Gann & Salter, 2000), although proponents of offsite or modular construction suggest that there are benefits in costs and quality from using more approaches derived from manufacturing. The project approach, delivered through teams with complementary skills and knowledge, leads to a high value being placed on social capital and trust in the supply chain (Wade et al., 2016). At the same time, the RMI elements of construction are trapped in a low skills low innovation equilibrium where increases in quality assurance are seen as a cost burden (Killip, 2020).

A sector will also manifest differently in different locations and contexts and it is important to scope the boundaries of the sector you're considering carefully. For construction, even a cursory comparison of the US and the UK shows immediate differences. The UK makes extensive use of blockwork while the US uses timber framing. Only about 12% of single-family homes in the US are under 1000 ft^2 (100 m^2) whereas only about a quarter of English homes are larger than that. In the US nearly two-thirds of homes are single-family detached whereas this characterises only about a quarter of homes in the UK. The UK also has older housing stock, with one in five homes having been built prior to 1919 compared to only 7% of the US housing stock (U.S. Census Bureau, 2019; Ministry of Housing, Communities and Local Government, 2020). There are also quite distinct space conditioning technologies, although both countries predominantly use natural gas for heating. The UK is dominated by boilers and radiators, and the majority of US homes use forced-air furnaces.

1.2   Step 2: Assess the sector drivers and interfaces

The second stage in your sector-specific research investigating is to ask "Why is it this way? What are the factors that lead to any given business sector's distinct features?" These motivations did not develop by accident, and any attempt to change practices must consider both what needs are being served and the systems that serve them.

Differences between the UK and US in housing age, size, and common materials will derive from a variety of historical and cultural factors. The UK is older than the US, has a long history of using block construction, and has had more centralisation of the government. The US is newer, has a pioneer and individualistic ethos, more decentralisation with power shifted to individual states and local municipalities, and a large supply of forests.

Climate explains differences in space conditioning. Radiators used to be common in US housing and can still be found in older homes. However, much of the US gets very hot in the summer and the advent of air conditioning provided people with a way to be comfortable in these locations. Since cooling systems were developed that used forced-air it became sensible to combine the two into a single duct system to distribute both heating and cooling throughout the home. The UK, on the other hand, does not have the same hot climates and so the move towards air conditioning and forced-air was less important.

The prevalence of forced-air systems in the US means that measures intended to provide acceptable indoor air quality often get integrated into the space conditioning systems. Filtration, which was originally incorporated into the systems to protect the equipment, has become a mechanism for removing particles in homes. In some cases, whole-dwelling ventilation gets integrated with the forced-air system. In the UK, indoor air quality measures largely remain separated from heating.

On the other hand, forced-air systems are more prone to contributing to energy losses. Because of their size they get put into unconditioned spaces such as attics. Also, whereas radiators that carry hot water cannot leak without serious damage to the building, air duct leaks do not carry the same level of hazard and so often are not made airtight. One outcome of this related to carbon management is that it is often not enough to improve the efficiency of the heating appliance – the ducts, which can be difficult or unpleasant to access, also need to be addressed.

Another major difference between the UK and US relates to the centre of governance for standards and codes. In the UK standards and codes are largely set at the national level. In the US, housing standards and codes are adopted at the local level. This results in substantial heterogeneity regarding requirements in different locations within the country. This also makes it more difficult to take a unified approach to energy and carbon reduction. One commonality between the US and the UK, and most developed countries, is that RMI construction has been happening for a long time and there are long-established practices. In most cases people do not learn construction by going to school but rather learn from experienced people who will mainly teach their own practices, making innovation very difficult. The delivery mechanisms for new understanding, approaches, and technologies need to take this learning pathway into account (Simpson & Owen, 2020).

1.3   Step 3: Understand existing research

Having gained an understanding of your sector's key characteristics and the forces that shape it, the third stage in developing sector-specific research is to increase your understanding of the existing research landscape in that sector. What research is, or is not, being undertaken? Why? What insights is that research providing, and is it built upon assumptions that you want to test? Is that research related – implicitly or explicitly – to sustainability?

In this context, identifying the main channels for publishing research, both peer reviewed and not, is arguably more important than a systematic review; you are trying to gain a broad understanding of what is happening in the sectoral research, rather than identifying research gaps at this stage. For

RMI construction, rapid review of international peer-reviewed journals such as *Buildings and Cities, Building Research & Information*, or *Construction Economics and Management* provide such sector insights. Such a review shows that the flow of innovative technology and design ideas to apply to existing buildings continues, with studies that evaluate the impact of such innovations taking place, typically at a small or case study scale. There are also strands of research discussing business models for innovation in RMI, how innovation links to social practices of home owners, the delivery of energy services in homes, skills challenges in the sector, impacts of housing on health and resource consumption, as well as research which seeks to understand and develop construction processes – such as design for performance and project management.

In RMI construction, as well as other fields, there are at least three major categories for research: technology/tools, market/demand influences, and impacts. Each of these can have multiple facets. For example, market/demand influences may include how social factors influence the need for and adoption of technology as well as how to get practitioners to begin implementing the technology. A non-exhaustive list of the areas of research in RMI construction includes:

- Technology/tools
  - Development of higher-efficiency equipment
  - Development of easier-to-apply air sealing techniques
  - Development of improved insulation
  - Development of improved insulating approaches
  - Development of easier/faster approaches to construction
  - Integration of components
  - Enhanced controls/communication
- Market/demand influences
  - Differential needs by social factors, e.g. socio-economic status, age, race, gender
  - Readiness for adoption by practitioners
  - Readiness for adoption by occupants/owners
  - Ease-of-use
  - Educational needs
- Impacts
  - Energy use
  - Indoor environmental quality
  - Carbon reduction
  - Rates of adoption
  - Economic impacts.

Energy savings and economic/carbon impacts are a major focus of residential research. In the US, technologies for new construction receive substantial funding from the US Department of Energy (USDOE) as the federal government aims to move new construction towards zero energy or carbon neutral housing. In the UK, there is also large-scale support for innovation in new building through the "Transforming Construction" programme supported by the UK Industrial Strategy, which focuses on modular and offsite construction methods as routes to higher quality low carbon homes.

Although new construction represents only about 1% of homes per year, existing homes do not receive as much research attention. USDOE has funded some studies on retrofit-ready technologies; on both sides of the Atlantic some retrofit programmes conduct impact evaluations that typically do not get published in the peer-reviewed literature; and economists have taken an interest in understanding the costs and benefits of energy efficiency primarily as a driver for addressing climate change through reducing carbon emissions associated with housing.

Meanwhile, there is an acknowledgement that the indoor environment can impact health (which has been amplified in 2020 in the face of a global pandemic), but indoor air quality and health have received less attention. One problem is "ownership". The US Department of Housing and Urban Development (USHUD) does fund research on "healthy homes", usually focusing on how programmes impact incidence rates of illness or how actions impact contaminant concentrations. Relative to other spending on building impacts, funding allocated to USHUD's programmes is small. The US Environmental Protection Agency (USEPA) has no regulatory authority over indoor air and so has little funding in this area outside of addressing radon, which is a contaminant that has also garnered substantial analysis in the UK. One area of inquiry that has received attention in housing and health is "social determinants of health", which aims to examine the relationship between health and social issues such as poverty and neighbourhood effects.

Contrasting the buildings-related research concerned with energy in buildings and internal air quality is instructive. The key difference and, we speculate, an important factor in the funding for and impact of research in these two areas lies in how visible and well understood are the links and logic that connect the observed phenomena with the sustainability outcome. Given the difference in funding it is ironic that adoption of energy efficiency is sometimes impeded by concerns about health.

There is no doubt that changes in internal air quality lead to health impacts, but these impacts are also dependent on a range of other factors such as the household's socio-economic status, house type and condition, family history, underlying condition as well as their habits and practices in the home. How all these factors relate, and the precise relationship between indoor air quality (IAQ) and health benefits, is complex. Therefore, it is more difficult to evaluate overall impacts or to predict impacts for an individual occupant. Whereas energy efficiency can be related to reduced energy use, reduced energy spending, and reduced carbon emissions, impacts on health and well-being will vary from person to person.

Energy consumption is also complex, but it is shaped by fewer related variables, mainly house type and condition, together with the demography of the household. In addition, there are clear legislative and policy drivers to reduce greenhouse gas emissions.

1.4   Step 4: Explore sustainability impacts

Having gained an understanding of the sector's characteristics, an appreciation of why the sector operates in the way that it does currently, and identifying why and how research gets done in the sector, the fourth stage in developing sector-specific research is to determine the opportunities to extend our knowledge regarding the sustainability impacts of that sector. This enables you to focus your research where it will have the most benefit – addressing a gap in knowledge and tackling a sustainability issue.

Any sector will have both direct and indirect impacts on sustainability. An impact can be positive or negative and it's worth taking some time to think through the best way of representing those impacts for your sector of interest. Sustainability is a complex idea and set of goals, so you need to take some positive decisions about how to structure the big idea so that it becomes tractable. The triple bottom line (Elkington, 1999) of social, economic and environmental costs and benefits is one, necessarily over-simplified, framework for thinking about these impacts. While the triple bottom line is critiqued for allowing "weak" interpretations of sustainability (Isil & Hernke, 2017) and facilitating trade-offs between different impacts, it can be useful because of its currency in business sustainability thinking; the very reasons that make academics wary of the framework – the analogy with accounting, the simplified categories and the search for metrics – are also the reasons why it is attractive to businesses finding their way in sustainability strategy. The "five capitals" (Porritt, 2007) approach to assessing sustainability adds some further dimensions but the same tensions between the appeal of a model which uses a financial lens and monetisation to understand value and the limitations of that model in representing messy reality mean that the use of "capitals" is also heavily critiqued in sustainability research.

For our example, we use the triple bottom line as an organising idea, recognising the risks but, in this case, judging that the simplicity of the framework and the way it facilitates communication with sector actors is our prime criterion. For construction businesses' activities, direct impacts include:

- Environmental impacts: resources used in creating the building or structure, waste generated during construction and maintenance, habitat creation/destruction, water management around the construction site, including containment of pollution.
- Economic impacts: employment and jobs created, creating buildings which are economic assets, changing/property land values.
- Social impacts: providing shelter and supporting well-being; skills development and employment.

Assessing the "social value" of construction projects in terms of how they offer benefits to excluded groups is a particularly active research topic, driven by both regulatory changes, related to finance and workforce policy, and by a greater understanding of the need for representation. In the UK, legislation (the Public Services (Social Value) Act 2012) means that public sector procurement demands that contractors demonstrate how they will generate social value. Social value is in turn responsive to the specific context for construction; for example, Australian projects are gathering evidence about the integration of indigenous peoples into construction projects (Denny-Smith et al., 2020); research in the UK explores the precarity of migrant workers with limited English language skills (Tutt et al., 2013).

Indirect impacts are equally diverse but, given the purpose of construction is (usually) to create long-lived structures, these impacts are felt over a longer period:

- Environmental – whole life energy consumption, impacts of resource extraction away from construction site, embodied energy e.g. the carbon intensity of steel.
- Economic – land and property prices, the contribution of infrastructure (transport, energy, information & communication technology [ICT]) to supporting economic activity and development.
- Social – sustainable communities, healthy living, reduced burden on under-resourced communities.

Considering two of these impacts in – slightly – more depth helps us to see how specific research topics within a sector can be shaped to make a broader contribution to sustainability.

First, the impacts of construction on health and the changing nature of research inquiries illustrate how a sustainability issue can morph over time, requiring different data collection and analysis. Health in construction was originally a concern for labour and employment practices, avoiding injury, or death of construction workers. The health impacts of construction are now equally concerned with how buildings impact the health of the people who use that building through design, use of materials, or incorporating biodiversity. Research has expanded to evaluate how these innovations impact upon building users through post-occupancy evaluation and modelling approaches to understand what might be achieved through industry-wide changes in practice.

Second, research on the sustainability of resource use in construction has developed from the starting point of understanding the impacts of materials used in construction and avoiding waste. There is an interesting parallel with the changing focus on construction and health impacts here, as the need to improve resource efficiency and reduce waste was codified in regulations in the UK in 2008 with a requirement for every construction site to develop a waste management plan and have clear lines of responsibility and accountability for compliance with that plan. However, data now shows that a building has much greater carbon emissions associated with energy consumption while the building is in use than in its operational phase (Ramesh et al., 2010) although improvements in energy efficiency are increasing the relative importance of carbon emissions embodied in construction materials (Röck et al., 2020). A further issue has surfaced from building sustainability research: the "design–performance gap" – the difference between the energy demand predicted by the designer's modelling software and what a building actually uses.

All these issues reinforce the need for new, sector-specific research that explores, understands and suggests solutions to sustainability challenges in a way which is readily seen to be relevant to the construction industry – both policymakers and practitioners.

1.5   Step 5: And repeat!

Defining research questions is rarely a linear process. In following these steps above, you will uncover new factors that you want to recognise in your research design as you characterise the sector in step 1. Step 2 may identify interfaces between actors or systems which you had not considered before, but which seem intriguing and mean that you want to change the boundaries of the sector you are researching, meaning that you have to revisit how you characterise it (Step 1). You may find that your original idea is simply not that original when you explore what has already been done in step 3. Or you may find that the sustainability impacts you identify in step 4 lead you to some quite different understandings of what you should prioritise. Use these four steps as a structure to guide you to a focused research approach, but do not be constrained by them.

Eventually you will feel comfortable with how you have refined your research scope and can focus on what your research has uncovered.

Returning to our contrasting research areas within construction RMI, what both energy and IAQ have in common is the paucity of empirical research that tracks and evidences, at scale, the connections between that topic and the desired sustainability outcome. In trying to bridge the gap between small empirical datasets and the major impacts needed, researchers often turn to models to scale up the available data. Unfortunately, models do not sufficiently consider the range of defects in buildings especially when based on small samples which may not be representative of the population, and so the insights they offer are rarely usable at the small-scale level of what individual RMI practitioners need to do. Thus we come to the first of our sector-specific suggestions for a sustainability research agenda: the need finds ways to close the gap between the pioneers and mainstream practice.

To research adequately, and make large-scale impact on, sustainability of the RMI construction sector requires more than academic inquiries. Solutions are not simply technological, and so the research cannot simply focus on the technology. Instead, it will need to be done in full collaboration with those whose practices can make or break attempts at advances, and with a focus on addressing their pinch points. This collaboration needs to be done at both the fundamental understanding stage and at the market transformation stage. This is our second sector-specific suggestion, that developing research must be done alongside (reflective) practitioners.

2.   Methods

Having followed these four steps for your own area of interest, and identified the aspect of sustainability in your sector that you wish to research in more depth, you need to use research methods which reflect the dimensions of the problem you are investigating.

Business sustainability research problems are interdisciplinary by nature, and "mixed methods" would appear to be an obvious route to follow.

It is easy to say that the methodology, and methods selected as part of that methodology, should be developed in direct response to the problem the research is intended to tackle. In practice, it is also easy to be swayed from that clear intent. Researchers have different capabilities and preferences for methods, often reflected in a choice for quantitative or qualitative methods rather than blending the two. Similarly, the accessibility of data will be an influence in research design; if a perfectly designed project relies on data which does not exist or cannot be collected, then that project is unlikely to be successful!

In our example, a substantial amount of research is being conducted on residential buildings. Three data challenges limit how far this research is based on empirical data. The practical challenge of accessing homes for data collection, combines with the technical methodological challenge of assessing the impact itself – whether that be carbon emissions or individual health – to make it difficult to collect datasets which do more than describe the challenge in greater detail. Added to this, the challenge of constructing datasets which offer representative sampling and scale also constrains the claims that can be made for research. For example, while qualitative data gathered from supply chain actors might reveal that construction materials wholesalers (called "builders merchants" in the UK) are critical to the decisions made on individual RMI projects (Killip et al., 2020), the data that underpins that assertion remains based on a small sample of practitioners, which limits the impact those research findings can have on policymakers, or on the broader industry.

As a result, construction industry research on sustainability topics often turns to modelling. Field monitoring may be conducted, but this is used to validate models rather than to derive new models or understanding. Models can indeed help to understand a problem, but using modelling to identify solutions and sustainability impacts is dependent on creating scenarios which the model can run. Assumptions have to be made in order to make a model "work", but those assumptions can be "heroic" in nature and can reduce the validity of model results. How does a model of training needs for construction workers yield useful information if it is based on a single "typical" construction worker, when the construction labour force is in fact hugely varied?

There are possibilities offered by new, agent-based, approaches to modelling which focus on individual behaviours, intertwined with a wide range of contextual factors and operating in a range of values, rather than a precise, deterministic value (Owen & Heppenstall, 2020). This type of approach is an example of the complexity of methods that sector-focused and problem-focused business sustainability research demands. This type of research requires interdisciplinary working, fully integrating physical, natural and social sciences. Research vocabularies for each discipline need to be expanded so that qualitative and quantitative data, empirical and modelled, reflecting inputs and impacts can be developed.

3.   Conclusion

What does our exploration of sustainability research in RMI construction mean for your sector-specific research agenda? Our example has some attributes which are typical for other sectors. Culture, history and geography (physical and human) have major influences on the fundamental form of the system that cannot be easily moved aside. Complex supply chains drive construction activity and are also a feature of many retail sectors. Many of the sustainability impacts of the sector lie beyond the immediate operations of businesses in the sector, but the size of the impact is determined by what the business does, with the energy demand of a building throughout its life set by the activities of the firms that create, repair or renovate the building. This is also true for automotive businesses. The sustainability impacts of construction materials on land use or labour standards have echoes in the concerns of textile and clothing production. But some of construction's attributes do not easily read across to other sectors. Construction, and particularly RMI construction, is dominated by a huge number of small firms, in sharp contrast to the concentration into a small number of large firms for, say, energy supply or food retail. This is why we find the step-wise approach to developing your research topic helpful; it ensures that you respond to the specific characteristics of your area of study.

For the construction RMI sector, our own research responds to two research priorities which emerge from following the four steps as outlined above. First, understanding the sector and its impacts leads to identifying a clear difference between best practice and what the average business in the sector does. There is a strong need for research which focuses on how to close the gap between the pioneers and mainstream practice. Second, we observe that a lot of research is based on models and technology-driven understandings of what's possible. This is insufficient to change practice and influence the immediate impacts of the construction sector. We need to develop research alongside (reflective) practitioners so that research insights are accurate, and useful, for the practitioners whose day-to-day activities are what shape building performance, and building sustainability.

We do not claim that these are the only research topics that can be undertaken to explore sustainability and the construction sector. There are myriads of interesting aspects of the sector where knowledge can be extended through research. However, our experience in using the step-by-step approach outlined above, has led us to research foci which give us the confidence that our investigations go beyond being interesting, or being descriptive, extending to being useful in helping the sector improve its sustainability performance.

References

Deloitte. 2019. Changing the lens: GDP from the industry viewpoint, *Economics Spotlight*, July 2019. https://www2.deloitte.com/us/en/insights/economy/spotlight/economics-insights-analysis-07-2019.html

Denny-Smith, G., Williams, M. & Loosemore, M. 2020. Assessing the impact of social procurement policies for Indigenous people, *Construction Management and Economics*, 38(12), 1139-1157.

Elkington, J. 1999. *Cannibals with Forks: Triple Bottom Line of 21st Century Business*, London: Capstone.

Emrath, P. 2018. More new homes needed to replace older stock. National Association of Home Builders. https:// www .nahbclassic .org/ generic .aspx ?sectionID = 734 & genericContentID=263243

Gann, D. & Salter, A.J. 2000. Innovation in project-based, service-enhanced firms: The construction of complex products and systems, *Research Policy*, 29(7-8), 955-972.

Isil, O. & Hernke, M.T. 2017. The triple bottom line: A critical review from a transdisciplinary perspective, *Business Strategy and Environment*, 26, 1235-1251.

Killip, G. 2020. A reform agenda for UK construction education and practice, *Buildings and Cities*, 1, 525-537.

Killip, G., Owen, A. & Topouzi, M. 2020. Exploring the practices and roles of UK construction manufacturers and merchants in relation to housing energy retrofit, *Journal of Cleaner Production*, 251, 119205.

Ministry of Housing, Communities and Local Government. 2020. English Housing Survey data on stock profile. https://www.gov.uk/government/statistical-data-sets/stock-profile

National Association of Home Builders. 2020. Housing's contribution to gross domestic product. https:// www .nahb .org/ News -and -Economics/ Housing -Economics/ Housings-Economic-Impact/Housings-Contribution-to-Gross-Domestic-Product

Office for National Statistics. 2019. Construction statistics annual tables. https:// www .ons .gov .uk/ busin essindustr yandtrade/ c onstructio nindustry/ datasets/ cons tructionstatisticsannualtables

Owen, A. 2015. Missing the point – the challenge of creating policies and programmes that tap into the motivations of the builders and installers, *ECEEE Summer Study*, Presqu'île de Giens, France.

Owen, A. & Heppenstall, A.J. 2020. Making the case for simulation: Unlocking carbon reduction through simulation of individual 'middle actor' behaviour, *Environment and Planning B: Urban Analytics and City Science*, 47(3), 457–472.

Porritt, J. 2007. *Capitalism as if the Earth Mattered*, London: Routledge.

Power, A. 2010. Housing and sustainability: Demolition or refurbishment? Proceedings of the Institution of Civil Engineers – *Urban Design and Planning*, 163(4), 205–216.

Ramesh, T., Prakash, R. & Shukla, K.K. 2010. Life cycle energy analysis of buildings: An overview, *Energy and Buildings*, 42, 1592–1600.

Rhodes, C. 2019. Construction Industry: Statistics and policy. House of Commons Library. Briefing paper 01432.

Röck, M., Saade, M.R.M., Balouktsi, M., Rasmussen, F.N., Birgisdottir, H., Frischknecht, R., Habert, G., Lützkendorf, T. & Passer, A. 2020. Embodied GHG emissions of buildings – the hidden challenge for effective climate change mitigation, *Applied Energy*, 258, 114107.

Simpson, K. & Owen, A. 2020. Reflections from engaging a network of local stakeholders in discussing training needs for sustainable renovation, *Frontiers in Built Environment*, 6, 142.

Statista. 2020. Number of production workers within the U.S. construction industry from 1998 to 2018. https://www.statista.com/statistics/193094/employment-in-production-within-us-construction-since-1996/

Tutt, D., Pink, S., Dainty, A.R.J. & Gibb, A. 2013. 'In the air' and below the horizon: Migrant workers in UK construction and the practice-based nature of learning and communicating OHS, *Construction Management and Economics*, 31(6), 515–527.

U.S. Census Bureau. 2019. American Housing Survey. https:// www .census .gov/ programs-surveys/ahs.html

U.S. Census Bureau. 2020. Monthly New Residential Construction, October 2020. https://www.census.gov/construction/nrc/pdf/newresconst.pdf

Wade, F., Shipworth, M. & Hitchings, R. 2016. Influencing the central heating technologies installed in homes: The role of social capital in supply chain networks, *Energy Policy*, 95, 52–60.

9   A research agenda for the extractive industries and the low carbon transition

Laura Smith, James Van Alstine and Alesia Ofori

1.   Introduction

There is an urgent need to move away from fossil fuels towards renewable sources of energy. The Intergovernmental Panel on Climate Change's (IPCC) recent damning report on the status of climate action warned that without immediate and deep emissions reductions across all sectors, the goal of limiting global warming to 1.5°C is beyond reach (IPCC, 2022). Climate change targets include a move from petrol and diesel to electric vehicles, a shift to wind and solar technologies, and improved energy storage, all of which require significant mineral and metal inputs.

While on the one hand the high demand for minerals and metals promises a new era of prosperity for populations with extractive economies, evidence suggests that the mining of metals and minerals needed for the low carbon transition is likely to continue the exploitative extractive relationships that characterise minerals extraction in many parts of the world, especially in the Global South. Scholars and non-governmental organisations (NGOs) note new and existing geographies of risk, including human rights abuses, gender-based violence, environmental destruction, and conflict (Bonds and Downey, 2012, Dunlap, 2021, Sovacool, 2021).

Further, because the future supply of minerals is closely tied to the sustainable development of the sector, it is argued that mining companies will have to demonstrate their commitment to Environment, Social and Governance (ESG) standards at a level previously unseen (Valenta et al., 2019). ESG in the mining sector has grown in recent years, with investors turning to sustainable investment criteria to evaluate companies that might pose financial risk due to their environmental or other practices. ESG follows a long and complex history of Corporate Social Responsibility (CSR) in the sector, which is heavily critiqued as failing to address the complex and deep-seated social issues that arise in extractives contexts. A recent review of ESG performance suggests a continuation of these challenges in relation to the sector's social and environmental performance (Responsible Mining Foundation, 2022).

Considering the mining intensity of the low carbon transition there is a clear need for a research agenda to flesh out what sustainable business means in this context. Many of the concerns raised about the low carbon transition are not new, but rather reflect long-standing issues around resource extraction and sustainable development in the Global South. There is, therefore, a need for a deeper understanding of existing and ongoing challenges around mining, and a link made to current thinking about business engagement with green technologies. Understanding the complexities will not only enable business to address impacts and engage with green technologies in ways that minimise harm, but in ways that seek to be more inclusive and more just.

In this chapter we explore change and continuity in mining for the low carbon transition, first revisiting literature on the transformational impacts of mining and then exploring the links between mining and sustainable development. We then go on to discuss the issue of conflict minerals and artisanal mining, both of which are playing a role in mining for the low carbon future. We then reflect on the challenges facing ESG in the sector in light of these challenges. We conclude the chapter by identifying a potential research agenda for the topic.

2. Mining and the low carbon transition: change and continuity

The mining sector is playing a crucial role in the transition away from fossil fuels towards low carbon technologies. The technologies required, including wind turbines, solar panels, electric vehicles and improved energy storage, need significant mineral and metal inputs, meaning that demand is set to accelerate considerably (Blondeel et al., 2021). Conservative estimates suggest that solar photovoltaic (PV) will account for 87 per cent of aluminium demand; wind and geothermal will account for 98 per cent and 64 per cent of zinc and titanium demand respectively; and 74.2 per cent of all copper demand will come from solar PV and wind combined (Hund et al., 2020). Graphite, lithium, and cobalt will need to be significantly ramped up by more than 450 per cent from 2018 levels by 2050 to meet demand from energy storage technologies (ibid.).

Estimates by NGOs are more severe. Bolger et al. (2021) argue that under a 'high demand' decarbonisation scenario, batteries for electric vehicles and renewables are likely to drive up demand for lithium, found in Latin America, by almost 6000 per cent by 2050. The global production of cobalt, found almost exclusively in the Democratic Republic of the Congo (DRC) in Africa, is expected to almost double by 2030 if it is to meet the ambitious targets for electric vehicles being set by governments around the world (Mancini et al., 2021). Valenta et al. (2019) predict that the demand for copper could increase 300 per cent above current levels, requiring extraction from new orebodies that are lower grade and deeper, and will therefore create a larger footprint.

Concerns are being raised about the implications of supplying the demands for these strategic metals and minerals, including challenges of scarcity, conflicts, and destructive mining practices creating similar problems to those that arise from extracting fossil fuels. Indeed, Temper et al. (2020) show that low carbon, renewable energy and mitigation projects are as conflictive as fossil fuels projects, and that both disproportionately impact vulnerable groups, such as rural communities and indigenous peoples. Citing similar concerns, Sovacool (2021) identified 61 different vulnerable groups of indigenous peoples, aboriginal collectives, or ethnic minorities negatively impacted by green technologies globally. In addition to human rights abuses, gender-based violence, environmental destruction, and conflict, there are new areas of risk, such as deep-sea mining, which lacks the governance frameworks to ensure the necessary environmental and social safeguards (Kung et al., 2021).

3.   Understanding mining's transformational impacts

For many communities, mining for the low carbon transition continues long and complex histories of mining, of promises of prosperity from extractive economies, and of dispossession and exploitation. Mining has considerable environmental and social impacts, transforming land and landscapes, and the lives and livelihoods of local and regional communities (Gamu et al., 2015). Mining often takes place in regions inhabited by politically marginalised populations experiencing ongoing challenges of poverty and exclusion that are further exacerbated by mining activities (Acuña, 2015, Buur et al., 2017). Mining companies become a dominant economic force in the region, violating the rights of local communities, buying up land, enclosing natural resources, altering access to a range of resources needed for livelihoods, and reducing access to potable water. Indigenous peoples are often disproportionately affected, given their reliance on land and resources that are susceptible to environmental degradation from mining activities (O'Faircheallaigh, 2013). Land dispossession and resettlement often has long-term implications and increases vulnerabilities of communities. Scholars note that dispossession also encompasses the dispossession of health, habitat, way of life, and gain from resources (Acuña, 2015, Nixon, 2011).

Local populations bear the heavy social and environmental costs of mining, while receiving very little of the wealth they generate (O'Faircheallaigh, 2013). Mining tends to operate in an enclave fashion, where capital is invested in spatially segregated mineral-extraction enclaves providing little in the way of productive linkages to the local economy or society (Ferguson, 2005). Recent work on enclaves has shown that extractive enclaves are not entirely separate but interact with the local context in complex ways, to mediate access to rights and services, which can entrench class differences (Enns and Bersaglio, 2015, Lesutis, 2019).

Political mobilisation among communities adversely affected by mining has led to increased resistance to mining projects, and an escalation of social conflicts (Andrews et al., 2017). Many of these struggles lead to dynamic geographical alliances connecting local communities with global communities of support (Brown and Spiegel, 2017). Yet, while there has been a dramatic increase of local resistance to extractive projects, scholars find different objectives, narratives and intensity of resistance across extraction sites (Conde and Le Billon, 2017). Scholars note that social conflict is not always about resistance to extraction per se. Bebbington et al. (2008) show that conflict can be as much about the embedded cultural values and meanings associated with land and nature as about the protection of land and livelihoods. Therefore conflict can be deeply rooted in a region's social history and can reflect wider struggles over territory and identity (Hinojosa et al., 2015).

Mining for low carbon technologies inevitably becomes entangled in these complex and long-standing struggles over territory, rights and identity, which in many cases have been shaped by histories of extraction. For example, lithium mining in Latin America is impacting vulnerable and disempowered communities leading to conflict and a rejection of industry in Argentina and Bolivia (Barandiarán, 2019, Walter and Wagner, 2021). Consultations with indigenous populations have not been open and inclusive. In Bolivia, legal compliance has been prioritised over protecting the vulnerable Uyuni salt flat's ecosystem (Sanchez-Lopez, 2019). Cobalt mining in DRC continues long-standing challenges of conflict, human-rights abuses and child labour in the mining sector. In the next section, we look at how the concept of sustainable mining and ideas around transparency have emerged which attempt to address some of these issues.

4. The extractive industries and sustainable development

Growing public pressure in the late 1990s for the mining industry to respond to its negative environmental and social impacts led to the emergence of governance initiatives in the sector. A group of the world's largest mining companies initiated the Mining Minerals and Sustainable Development (MMSD) project to explore how the sector could contribute to sustainable development and address the industry's poor reputation (Buxton, 2012). As an outcome of this project, the International Council of Mining and Metals (ICMM) was created in 2001 which seeks to mainstream sustainable development and good ESG norms and standards across the mining and metals industry. CSR became the vehicle through which mining companies committed to maximise the sector's contribution to sustainable development (Buxton, 2012).

Around the same time as the industry-led MMSD project, 'ideas' on transparency and accountability for the extractive industries emerged from the intersection of international networks concerned with corruption, conflict and CSR (Van Alstine, 2014). NGO advocacy drove the international transparency agenda – for example, an important catalyst was Global Witness's 1999 publication 'A Crude Awakening' highlighting the role of the oil and banking industries in Angola's civil war and the 'hollowing out' of the state.

The Extractive Industries Transparency Initiative (EITI), established in 2003, has become the transparency and accountability standard for the oil, gas and mining sectors. EITI is a voluntary mechanism which aims to facilitate the management of revenues and ensure accountability through company reporting of the payments they make to governments, and governments reporting the amounts they receive from companies (Van Alstine and Smith, 2018). Unique to this multi-stakeholder initiative is that, although championed by NGOs and industry, it is implemented by governments. More recently, the EITI has evolved into an industry standard that seeks transparency and accountability across the extractive industries value chain.

While there have been some notable achievements through EITI, including procedures put in place in host countries for multi-stakeholder consultations and oversight (Van Alstine, 2014), critics highlight that the EITI fails as an accountability tool. A key reason is that the EITI often appears to be a 'tick-box' exercise that focuses narrowly on transparency, and rarely delivers real impact on the ground (Klein, 2017, Van Alstine, 2017). Nonetheless, transparency initiatives such as the EITI and mining companies' efforts towards 'sustainable mining' through CSR, are key governance mechanisms in the sector that emerged from pressures for change (Szablowski and Campbell, 2019).

The idea of sustainable mining was seen by many as a contradiction, given the extraction of non-renewables and mining's heavy social and environmental footprint. From this perspective, aside from keeping the minerals in the ground, the best the industry could do was to minimise any harmful impacts and improve resource efficiency (Brereton, 2003). An alternative perspective suggested the potential for mining to contribute to sustainable development. In a 2000 research paper, Hilson and Murck (2000) argued that mining companies could considerably improve their sustainability performance at the site level, if there was a commitment to continuous environmental and socio-economic improvement throughout the entire project lifecycle, from mineral exploration, through operation and to mine closure. However, while the MMSD project committed to sustainable development in principle, two decades on and scholars continue to highlight the disappointing outcomes. Buxton (2012), assessing the MMSD project ten years on, found that good intentions at the strategy level had not translated to progress on the ground. More recently, Kemp and Owen (2018) note the continuing problem of mining companies failing to account for the complexities of the local context.

Perhaps unsurprisingly, given the ongoing disconnect between strategy and action on the ground, social conflict remains one of the key ESG risks facing the sector. A study by Andrews et al. (2017) found that social conflict in relation to extractive industries increased dramatically in the period 2000 to 2013. Moreover, there is clear evidence that mining conflict is also related to low carbon minerals and metals. In Latin America, The Observatory of Mining Conflicts of Latin America (OCMAL) registered 284 social conflicts relating to mining, mainly in Mexico, Chile and Peru, followed by Argentina, Brazil and Colombia (OCMAL, 2022). Some of these conflicts are directly linked to strategic minerals. The Global Environmental Justice Atlas lists 323 socio-environmental conflicts related to the extraction of copper, lithium, rare earth elements and silver (Bolger et al., 2021). What continues to confound the mining industry is that social conflict in the sector continues to increase alongside the efforts of enlightened companies and their supporters (e.g. IFIs and industry associations) to develop best practice guidelines and toolkits to improve social performance (Franks et al., 2014, Larsen, 2015). This seeming paradox suggests that the dynamics of mining and social conflict continue to be poorly understood.

5.   Conflict minerals and green supply chains

A further concern relating to conflict and low carbon technologies is that 'conflict minerals' are entering so-called green mineral supply chains (Church and Crawford, 2018). Conflict minerals are minerals that are linked to violence, conflict and human rights abuses. For example, the production and sourcing of minerals such as cassiterite (the ore for Tin), coltan (ore for Tantalum), wolframite (Tungsten ore), and gold (known as 3TG) often are controlled by groups, including state security forces and armed groups as a means to pursue violent objectives and fund wars and insurrection. This has led to the categorisation of the 3TG minerals as conflict minerals. These minerals remain critical in the electronics, aviation, medical, defence, software, and automotive industries (Barume et al., 2016). The largest percentage of 3TG is sourced from the Great Lake Region of Africa (Rwanda, Burundi and DRC), the latter being the chief hotspot of mineral conflicts. In the Eastern portion of the DRC for instance, 'rebel' groups have established quasi-dictatorships and exclusive authority over mining sites.

In addition to 3TG, the demand and sourcing of minerals for the low carbon future are also contributing to fragility, conflict and violence in some mineral rich countries. Cobalt, an essential mineral in the low carbon transition is known as the 'blood diamond of this decade', due to its association with violence in the DRC. Nickel, for solar panels and energy storage, is connected to murder, sexual violence and forced displacement in Guatemala (Church and Crawford, 2020).

The conflict mineral challenge is propagated by high demand and rising prices for consumer electronics. Factors such as the ease of transporting minerals from illicit mining sites to formal channels, state weakness and numerous buyers in complex trading chains further exacerbates the problem (Young, 2018). Companies located 'upstream' (i.e. from mine to smelter) and 'downstream' (i.e. from smelter to end-user) of the supply chain are therefore at risk of using conflict minerals. Global standards and certifications (e.g. the Conflict Minerals Certification Schemes (CMCS), the United States Dodd–Frank Act of 2010, the Analytical Fingerprint (AFP) method) require companies to prove that minerals are either not from conflict-affected areas or that production and trade have no affiliation to conflict financing and human rights abuses.

Calls have been made to extend the certification needed for conflict minerals to what has now been classified as 'green conflict minerals'. However, not much progress has been achieved with the 3TGs, indicating the gaps, limitations and politics surrounding these international certification processes (Le Billon and Spiegel, 2021).

6.   Artisanal and small-scale mining and its role in the low carbon future

A further challenge to green technology supply chains is the growth of Artisanal and small-scale mining (ASM) in sourcing and producing strategic minerals. ASM refers to mining operations with simplified forms of exploration, extraction, processing, and transportation. ASM requires low capital input and high labour input, and is often carried out by individuals, families, or cooperatives (OECD, 2016). ASM plays a major role in the global mining sector and is expected to make a significant contribution to the low carbon transition (Sovacool et al., 2021). In 2012, estimates suggested that the ASM sector was producing 15–20 per cent of global minerals and metals (Sidorenko et al., 2020), including base metals and ore minerals used for modern electronic devices, and copper, cobalt and other critical metals used for low carbon technologies (Haan et al., 2020). Furthermore, ASM is an important livelihood strategy in many mineral economies, directly employing at least 44.75 million people and providing support to 134 million people in related sectors (World Economic Forum, 2020). However, 80–90 per cent of those employed remain in the informal (unregistered and unregulated) sector (ibid.).

Despite the ASM's contribution to the low carbon transition, the largely informal nature of the sector is perceived as a challenge to socio-economic development due to issues such as tax evasion, land degradation, water pollution, child labour, and poor human rights conditions. In some countries, the sector is alleged to be connected to civil wars, terrorist activities, money laundering, and smuggling (Le Billon, 2008, Bryceson and Jønsson, 2013). Therefore, whilst ASM is important to the low carbon transition, it poses a considerable challenge for ideas of green supply chains.

Accordingly, there has been growing momentum towards the formalisation of the ASM sector, driven by the state, NGOs, major mining companies, and donor agencies. Formalisation seeks to integrate informal artisanal mining into the formal economy and make ASM activities socio-environmentally responsible (Schilling et al., 2020). Formalisation aims to secure livelihoods of

ASM miners and ensure a reduction in the social problems that accompany informal ASM activities (Wall, 2010). Formalisation also entails strengthening formal institutions to guarantee property rights for artisanal miners and simplifying administrative procedures to encourage informal miners to transition to formal mining (Chimhowu and Woodhouse, 2008, Spiegel, 2012).

However, formalisation in especially developing mineral economies tends to be techno-centric, top-down and imposed. Furthermore, it can conceal political motives that seek to retain control over ASM and enable the government to demarcate land to multinational mining companies rather than genuinely improve conditions for ASM miners (Ayelazuno and Mawuko-Yevugah, 2019, Hilson and Maconachie, 2020, Ofori et al., 2021). This leads to counterproductive effects, encouraging artisanal miners to illegally produce and trade minerals that risk entering green mineral supply chains.

7. CSR, ESG and the low carbon transition

The complexities and challenges associated with mining for the low carbon transition suggest ever-increasing scrutiny of mining companies' ESG credentials (Valenta et al., 2019). This area is not new for the mining sector, and follows an evolving agenda of CSR.

CSR in the extractive industries can be understood as a set of practices and discourses that emerged as a business approach for addressing a company's social and environmental impacts (Dolan and Rajak, 2016, Frynas, 2009). In the mining industry, CSR has gone through an evolution from first- and second-generation approaches to the more recent incarnation of CSR as ESG. A range of industry standards and guidelines emerged to guide companies to improve their social and environmental performance, including IFC performance standards, the Equator Principles, the Voluntary Principles on Security and Human Rights, the Global Reporting Initiative (GRI) and the concept of Free, Prior and Informed Consent (FPIC). Essentially, CSR programmes serve the purpose to ensure legitimacy externally, by demonstrating to shareholders and concerned stakeholders that companies are responsible global citizens, and internally to help secure the presence of mining companies locally (Billo, 2015). However, the record of CSR is poor, with studies showing its tendency to create divisions among communities, foster dependency on mining companies, and exacerbate social conflict (Gilberthorpe and Banks, 2012). Given that companies' motivations for implementing CSR are for reputation management and to ensure local acceptance, benefits to communities tend to be incidental and accountability measures are lacking (Smith et al., 2018).

ESG in part responds to CSR's accountability problem by establishing the requirement for companies to measure and report on environmental, social and governance criteria. Investors are aware that ESG failures can lead to major disruptions and significant losses as a result, therefore demands have increased for improved performance, a greater degree of transparency and reporting of mining companies' ESG risk exposure. However, a recent report by the Responsible Mining Foundation (2022) found that while formal ESG commitments are becoming the norm, there remains a considerable gap between stated commitment and positive impacts at site level. Large mining companies are performing poorly on issues that have long been established as crucial to maintaining local acceptance, known as the 'social licence to operate'. These include around tracking and improving the quality of community–company relations, addressing gendered impacts and health-related impacts of mining, and ensuring workers' salaries meet the living wage (Responsible Mining Foundation, 2022). Given the track record of CSR in the mining sector thus far, the concerns raised about ESG suggest an ongoing disconnect between sustainable development strategies of mining companies and action on the ground. Scholars have begun to argue that ESG is mainly used as a method for mitigating risk and as a stock selection screen, rather than a tool for additional value creation (Przychodzen et al., 2016, Van Duuren et al., 2016).

8.   Research agenda

Minerals for the low carbon transition will exacerbate existing and introduce new complex challenges facing the mining sector. Building on the discussion in this chapter, this final section sets out a research agenda on the topic.

8.1   Developing justice-focused ESG approaches

There are strong linkages between transparency and accountability initiatives and emerging ESG metrics. The framing of 'green conflict minerals' may push more renewable and green energy sectors (e.g. offshore wind, solar, battery storage, EVs) to engage meaningfully with reporting their ESG supply chain risks. However, a focus on risk above ethics means that the disconnect highlighted in the 2022 Responsible Mining Review will persist. Currently, adherence to international standards of conduct is closely related to managing risk (both to reputation and to operations) and minimising negative effects on the shareholder value upon which mining companies depend (Witte, 2018). Issues that are highly salient to the wellbeing of mining-affected people are not sufficiently addressed by external requirements or reporting frameworks. An important area of investigation, therefore, is the ways in which reporting frameworks can be developed to reflect the realities of local contexts, ensuring that they are not merely tick-box exercises or screening processes. It is essential to explore the ways in which mining for the low carbon transition intersects with ongoing struggles over livelihoods, rights, conflict, and social and environmental justice. This will enable a deeper understanding of local complexities and the development of evidence-based approaches to ESG that are justice-focused rather than business-centred.

8.2   Strengthening standards and certification schemes

Several voluntary transparency, accountability initiatives, mandatory extra-territorial regulations (e.g. Dodd–Frank) and conflict-free green mineral supply certification schemes (e.g. Conflict Mineral Certification Schemes) have emerged on mineral revenue and the value chain (e.g. EITI). A viable research avenue is to consider how standards such as EITI can act as a platform for resource-rich countries to navigate low carbon transition pathways. For example, a recent EITI policy brief argues that EITI reported data can serve as an entry point for debate and policymaking related to the energy transition pathways, as well as addressing the risks and opportunities related to critical minerals production (EITI, 2021). While improvements in openness of information are important, an increase in information does not mean that stakeholders can access the information or, if they can access it, are able to act on it (Acosta, 2013). In some contexts, stakeholders such as NGOs and mining-affected communities face restrictive and repressive conditions. Therefore, it is critical to consider how the gaps, limitations and politics of standards and certification schemes can be addressed to better inform and empower, and to improve resource governance for the low carbon transition.

8.3   Investigating appropriate formalisation policies for ASM

The high demand for low carbon minerals will be supplemented by the active work of informal artisanal miners across developing mineral economies. This will make more visible the challenges of regulating informality and the blurred lines between the formal and informal governance institutions. The implementation of formalisation programmes will become an arena of politics influencing the processes and outcomes of formalisation efforts. Formalisation must be designed and implemented by considering the complex broader socio-political environment of the low carbon political economy and the heterogeneous internal and external actors with competing interests that pose challenges to innovative change within the extractives sector. An important area for research is to explore the existing tensions, complexities, and contradictions associated with the work of artisanal miners. Further, critical engagement with the complex state–society interactions at all levels is needed to gain a deeper understanding of how power dynamics between actors might impede formalisation objectives.

8.4   Moving beyond business-focused best practice frameworks

Social conflict around mining extraction is likely to continue to take place alongside the growth of sustainability indicators, transparency and accountability standards and certifications, ESG reporting and social investment. While it is important that the industry continues to refine social performance guidelines, there is a need for approaches to be less 'business case' focused, and more sensitive to the complexities of local contexts and the role of the sector in driving social conflict. The key issue at stake is not the geographical concentration of these minerals and metals in 'corrupt and autocratic jurisdictions' as suggested by some commentators (e.g. Meiners and Morriss, 2022). Rather, the issue is the location of minerals and metals in contexts where existing vulnerabilities are likely to be further exacerbated by mining. As Bebbington (2014) reminds us, social protest and resistance to mining are important drivers of industry change and do not necessarily mean that 'best practice' is not working. What is important is that social protest is understood not only as a challenge of governance, but in its full complexities as "defensive responses to accumulation through dispossession...of land, territory, landscape and natural resources; property, self-governance, citizenship and cultural rights; and of the value inherent within the subsoil" (p.903). Researching these issues through an environmental and/or social justice-lens is important to move away from the business-centred focus that informs most best practice frameworks.

The research agenda outlined in the sections above requires context-specific research that draws across disciplines and has a justice focus. Drawing from disciplines such as political ecology, peace studies and anthropology can provide insights into power relations and the dynamic interactions between nature and society. The emphasis should be on interdisciplinary data collection that seeks to explore the meanings, opinions and lived experiences of those individuals impacted by the low carbon transition. Additionally, attention should be paid to methodologies that can grasp the challenges related to the complexities of society's transition to the low carbon future; a focus on the interplay of agency, structures, materiality, discourse, social interaction, place and time (Bhaskar et al., 2017). The challenge ahead is how to balance the urgency of the low carbon transition with ensuring the rights and dignity of people on the frontlines of mining for green technologies.

References

Acosta, A. 2013. Extractivism And Neoextractivism: Two Sides Of The Same Curse. In: Lang, M. and Mokrani, D. (eds) *Beyond Development: Alternative Visions From Latin America*. Rosa Luxemburg Foundation and Transnational Institute, pp.61–86.

Acuña, R. M. 2015. The Politics Of Extractive Governance: Indigenous Peoples And Socio-Environmental Conflicts. *The Extractive Industries And Society*, 2, 85–92.

Andrews, T., Elizalde, B., Le Billon, P., Oh, C. H., Reyes, D. and Thomson, I. 2017. The Rise In Conflict Associated With Mining Operations: What Lies Beneath. Canadian International Resources Development Institute, 1–127.

Ayelazuno, J.A. and Mawuko-Yevugah, Lord 2019. Large-Scale Mining And Ecological Imperialism In Africa: The Politics Of Mining And Conservation Of The Ecology In Ghana. *Journal of Political Ecology*, 26(1), 243–262.

Barandiarán, J. 2019. Lithium And Development Imaginaries In Chile, Argentina And Bolivia. *World Development*, 113, 381–391.

Barume, B., Naeher, U., Ruppen, D. and Schütte, P. 2016. Conflict Minerals (3TG): Mining Production, Applications And Recycling. *Current Opinion in Green and Sustainable Chemistry*, 1, 8–12.

Bebbington, A. J. 2014. Socio-Environmental Conflict: An Opportunity For Mining Companies. *Journal Of Cleaner Production*, 84, 34.

Bebbington, A., Hinojosa, L., Bebbington, D. H., Burneo, M. L. and Warnaars, X. 2008. Contention And Ambiguity: Mining And The Possibilities Of Development. *Development And Change*, 39, 887–914.

Bhaskar, R., Danermark, B. and Price, L. 2017. *Interdisciplinarity And Wellbeing: A Critical Realist General Theory Of Interdisciplinarity*. Taylor and Francis.

Billo, E. 2015. Sovereignty And Subterranean Resources: An Institutional Ethnography Of Repsol's Corporate Social Responsibility Programs In Ecuador. *Geoforum*, 59, 268–277.

Blondeel, M., Bradshaw, M. J., Bridge, G. and Kuzemko, C. 2021. The Geopolitics Of Energy System Transformation: A Review. *Geography Compass*, 15, E12580.

Bolger, M., Marin, D., Tofighi-Niaki, A. and Seelmann, L. 2021. Green Mining Myth, Friends of the Earth [Online] [Accessed 29 April 2022] https://friendsoftheearth.eu/wp-content/uploads/2021/10/Green-mining-myth-report.pdf

Bonds, E. and Downey, L. 2012. Green Technology And Ecologically Unequal Exchange: The Environmental And Social Consequences Of Ecological Modernization In The World-System. *Journal of World-Systems Research*, 18(2), 167–186.

Brereton, D. 2003. Promoting Sustainable Development In The Minerals Industry: A Multi-Disciplinary Approach. Annual Conference Of The Australasian Association For Engineering Education, Melbourne.

Brown, B. and Spiegel, S. J. 2017. Resisting Coal: Hydrocarbon Politics And Assemblages Of Protest In The UK And Indonesia. *Geoforum*, 85, 101–111.

Bryceson, D.F. and Jønsson, J.B., 2013. Mineralizing Africa And Artisanal Mining's Democratizing Influence. In: Bryceson, D.F., Fisher, E., Jonsson, B. and Mwaipopo, R. (eds) *Mining and Social Transformation in Africa*. Routledge, pp.15-36.

Buur, L., Nystrand, M. and Pedersen, R. H. 2017. The Political Economy Of Land And Natural Resources In Africa: An Analytical Framework, DIIS Working Paper.

Buxton, A. 2012. MMSD+10: Reflecting On A Decade Of Mining And Sustainable Development. IIED Sustainable Markets Discussion Paper, International Institute For Environment And Development (IIED), London, June.

Chimhowu, A. and Woodhouse, P. 2008. Communal Tenure And Rural Poverty: Land Transactions In Svosve Communal Area, Zimbabwe. *Development and Change*, 39(2), 285-308.

Church, C. and Crawford, A. 2018. Green Conflict Minerals, International Institute For Sustainable Development.

Church, C. and Crawford, A. 2020. Minerals And The Metals For The Energy Transition: Exploring The Conflict Implications For Mineral-Rich, Fragile States. In: Hafner, M. and Tagliapietra, S. (eds) *The Geopolitics Of The Global Energy Transition*. Springer, pp.279-304.

Conde, M. and Le Billon, P. 2017. Why Do Some Communities Resist Mining Projects While Others Do Not? *The Extractive Industries And Society*, 4, 681-697.

Dunlap, A. 2021. Renewable Energy And The War Of Progress. *The Peace Chronicle*, 13, 42-51.

EITI 2021. EITI Brief: Preparing for the energy transition [Online] October 2021 [Accessed 8 May 2022] https://eiti.org/documents/preparing-energy-transition

Enns, C. and Bersaglio, B. 2015. Enclave Oil Development And The Rearticulation Of Citizenship In Turkana, Kenya: Exploring 'Crude Citizenship'. *Geoforum*, 67, 78-88.

Ferguson, J. 2005. Seeing Like An Oil Company: Space, Security, And Global Capital In Neoliberal Africa. *American Anthropologist*, 107, 377-382.

Franks, D. M., Davis, R., Bebbington, A. J., Ali, S. H., Kemp, D. and Scurrah, M. 2014. Conflict Translates Environmental And Social Risk Into Business Costs. *Proceedings of the National Academy of Sciences*, 111, 7576-7581.

Frynas, J. G. 2009. Corporate Social Responsibility In The Oil And Gas Sector. *The Journal Of World Energy Law & Business*, 2, 178-195.

Gamu, J., Le Billon, P. and Spiegel, S. 2015. Extractive Industries And Poverty: A Review Of Recent Findings And Linkage Mechanisms. *The Extractive Industries And Society*, 2, 162-176.

Gilberthorpe, E. and Banks, G. 2012. Development On Whose Terms?: CSR Discourse And Social Realities In Papua New Guinea's Extractive Industries Sector. *Resources Policy*, 37, 185-193.

Haan, J. De, Dales, K. and McQuilken, J. 2020. Mapping Artisanal And Small-Scale Mining To The Sustainable Development Goals. *Newark DE: University of Delaware (Minerals, Materials and Society program in partnership with PACT)* [Accessed 12 October 2022] https://www.pactworld.org/Mapping%20artisanal%20and%20small -scale%20mining%20to%20the%20sustainable%20development%20goals

Hilson, G. and Maconachie, R. 2020. For The Environment: An Assessment Of Recent Military Intervention In Informal Gold Mining Communities In Ghana. *Land Use Policy*, 96, 104706.

Hilson, G. and Murck, B. 2000. Sustainable Development In The Mining Industry: Clarifying The Corporate Perspective. *Resources Policy*, 26, 227-238.

Hinojosa, L., Bebbington, A., Cortez, G., Chumacero, J. P., Bebbington, D. H. and Hennermann, K. 2015. Gas And Development: Rural Territorial Dynamics In Tarija, Bolivia. *World Development*, 73, 105–117.

Hund, K., La Porta, D., Fabregas, T. P., Laing, T. and Drexhage, J. 2020. *Minerals For Climate Action: The Mineral Intensity Of The Clean Energy Transition*. World Bank.

IPCC 2022. IPCC Sixth Assessment Report, Mitigation of Climate Change [online] 4 April 2022 [Accessed 8 May 2022] https://www.ipcc.ch/report/ar6/wg3/

Kemp, D. and Owen, J. R. 2018. The Industrial Ethic, Corporate Refusal And The Demise Of The Social Function In Mining. *Sustainable Development*, 26, 491–500.

Klein, A. 2017. Pioneering Extractive Sector Transparency. A PWYP Perspective On 15 Years Of EITI. *The Extractive Industries and Society*, 4, 771–774.

Kung, A., Svobodova, K., Lèbre, E., Valenta, R., Kemp, D. and Owen, J. R. 2021. Governing Deep Sea Mining In The Face Of Uncertainty. *Journal of Environmental Management*, 279, 111593.

Larsen, P. 2015. *Post-Frontier Resource Governance: Indigenous Rights, Extraction And Conservation In The Peruvian Amazon*. Springer.

Le Billon, P. 2008. Diamond Wars? Conflict Diamonds And Geographies Of Resource Wars. *Annals of the Association of American Geographers*, 98(2), 345–372.

Le Billon, P. and Spiegel, S. 2021. Cleaning Mineral Supply Chains? Political Economies Of Exploitation And Hidden Costs Of Technical Fixes. https://doi.org/10.1080/09692290.2021.1899959

Lesutis, G. 2019. Spaces Of Extraction And Suffering: Neoliberal Enclave And Dispossession In Tete, Mozambique. *Geoforum*, 102, 116–125.

Mancini, L., Eslava, N. A., Traverso, M. and Mathieux, F. 2021. Assessing Impacts Of Responsible Sourcing Initiatives For Cobalt: Insights From A Case Study. *Resources Policy*, 71, 102015.

Meiners, R. E. and Morriss, A. P. 2022. Addressing Green Energy's 'Resource Curse'. SSRN 4058763.

Nixon, R. 2011. *Slow Violence And The Environmentalism Of The Poor*. Harvard University Press.

OCMAL 2022. Conflictos Mineros en América Latina [Online] [Accessed 11 May 2022] https://mapaconflictosmineros.net/ocmal_db-v2/

OECD 2016. *OECD Due Diligence Guidance for Responsible Supply Chains of Minerals from Conflict-Affected and High-Risk Areas*. Second Edition.

O'Faircheallaigh, C. 2013. Extractive Industries And Indigenous Peoples: A Changing Dynamic? *Journal of Rural Studies*, 30, 20–30.

Ofori, A. D., Mdee, A. and Van Alstine, J. 2021. Politics On Display: The Realities Of Artisanal Mining Formalisation In Ghana. *The Extractive Industries and Society*, p.101014.

Przychodzen, J., Gómez-Bezares, F., Przychodzen, W. and Larreina, M. 2016. ESG Issues Among Fund Managers: Factors And Motives. *Sustainability*, 8, 1078.

Responsible Mining Foundation 2022. RMI Report 2022 [Online] [Accessed 8 May 2022] https://2022.responsibleminingindex.org/resources/RMI_Report_2022-Summary_EN.pdf#:~:text=The%20RMI%20Report%202022%20is%20an%20evidence-based%20assessment,results%20and%20extracts%20from%20the%20RMI%20Report%202022

Sanchez-Lopez, D. 2019. Sustainable Governance Of Strategic Minerals: Post-Neoliberalism And Lithium In Bolivia. *Environment: Science And Policy For Sustainable Development*, 61, 18–30.

Schilling, J., Schilling-Vacaflor, A., Flemmer, R. and Froese, R. 2020. A Political Ecology Perspective On Resource Extraction And Human Security In Kenya, Bolivia And Peru. *The Extractive Industries and Society*, 8(4), 100826.

Sidorenko, O., Sairinen, R. and Moore, K. 2020. Rethinking The Concept Of Small-Scale Mining For Technologically Advanced Raw Materials Production. *Resources Policy*, 68, 101712.

Smith, L., Tallontire, A. and Van Alstine, J. 2018. Development And The Private Sector: The Challenge Of Extractives-Led Development In Uganda. In: Fagan, G. H. and Munck, R. (eds) *Handbook On Development And Social Change*. Cheltenham, UK and Northampton, MA, USA: Edward Elgar Publishing, pp.69–88.

Sovacool, B. K. 2021. Who Are The Victims Of Low-Carbon Transitions? Towards A Political Ecology Of Climate Change Mitigation. *Energy Research & Social Science*, 73, 101916.

Sovacool, B. K., Turnheim, B., Hook, A., Brock, A. and Martiskainen, M. 2021. Dispossessed By Decarbonisation: Reducing Vulnerability, Injustice, And Inequality In The Lived Experience Of Low-Carbon Pathways. *World Development*, 137, 105116.

Spiegel, S. J. 2012. Formalisation Policies, Informal Resource Sectors And The De-/Re-Centralisation Of Power: Geographies Of Inequality In Africa And Asia. [Online]. Bogor, Indonesia. http:// www .cifor .org/ fileadmin/ subsites/ proformal/ PDF/RSpiegel1212.pdf.

Szablowski, D. and Campbell, B. 2019. Struggles Over Extractive Governance: Power, Discourse, Violence, And Legality. *The Extractive Industries and Society*, 6, 635–641.

Temper, L., Avila, S., Del Bene, D., Gobby, J., Kosoy, N., Le Billon, P., Martinez-Alier, J., Perkins, P., Roy, B. and Scheidel, A. 2020. Movements Shaping Climate Futures: A Systematic Mapping Of Protests Against Fossil Fuel And Low-Carbon Energy Projects. *Environmental Research Letters*, 15, 123004.

Valenta, R., Kemp, D., Owen, J., Corder, G. and Lèbre, É. 2019. Re-Thinking Complex Orebodies: Consequences For The Future World Supply Of Copper. *Journal of Cleaner Production*, 220, 816–826.

Van Alstine, J. 2014. Transparency In Resource Governance: The Pitfalls And Potential Of 'New Oil' In Sub-Saharan Africa. *Global Environmental Politics*, 14, 20–39.

Van Alstine, J. 2017. Critical Reflections On 15 Years Of The Extractive Industries Transparency Initiative (EITI). *The Extractive Industries and Society*, 4, 766–770.

Van Alstine, J. and Smith, L. 2018. *The EITI And Fair Taxation: Exploring The Linkages. Business, Civil Society And The 'New' Politics Of Corporate Tax Justice*. Cheltenham, UK and Northampton, MA, USA: Edward Elgar Publishing.

Van Duuren, E., Plantinga, A. and Scholtens, B. 2016. ESG Integration And The Investment Management Process: Fundamental Investing Reinvented. *Journal of Business Ethics*, 138, 525–533.

Wall, E. 2010. Working Together: How Large-Scale Mining Can Engage With Artisanal And Small-Scale Miners. *International Council on Mining and Metals (ICMM), London, Washington, DC: Oil, gas and Mining Sustainable Community Development Fund (IFC CommDev)*. [Accessed 12 October 2022] https:// www .commdev .org/ publications/ working-together-how-large-scale-mining-can-engage-with-artisanal -and-small-scale-miners/

Walter, M. and Wagner, L. 2021. Mining Struggles In Argentina. The Keys Of A Successful Story Of Mobilisation. *The Extractive Industries and Society*, 8, 100940.

Witte, A. 2018. *An Uncertain Future-Anticipating Oil In Uganda*, Göttinger Reihe zur Ethnologie – Göttingen Series in Social and Cultural Anthropology; 11

Universitätsverlag Göttingen. [Accessed 12 October 2022] https:// univerlag .uni -goettingen.de/handle/3/isbn-978-3-86395-360-7
World Economic Forum 2020. Making Mining Safe And Fair: Artisanal Cobalt Extraction In The Democratic Republic Of The Congo (September), 1–28.
Young, S. B. 2018. Responsible Sourcing Of Metals: Certification Approaches For Conflict Minerals And Conflict-Free Metals. *International Journal of Life Cycle Assessment*, 23(7), 1429–1447.

10 Business sustainability in SMEs: towards an Afrocentric research agenda

Samuel Howard Quartey

1. Introduction

The World Business Council for Sustainable Development explains business sustainability as a form of progress where businesses have greater opportunity and responsibility to support the growing needs of current and future generations without destroying the natural environment and ecosystems (Schmidheiny, 1992). Small and medium-sized enterprises (SMEs) play an instrumental role in the socio-economic development of African countries. According to Oguntoye and Evans (2017), SMEs are one of the most effective solutions to the continent's socio-economic inequalities. In Africa, SMEs are considered as drivers of the economy since they promote trade and commerce (Kamunge et al., 2014). They provide jobs and livelihood for people who would otherwise have remained unemployed and disengaged (Sampath, 2016). The rural nature of these businesses contribute to the revival of rural and regional communities, and social services and amenities are provided, which otherwise would not have existed.

Despite their contributions towards sustainable development, SMEs in Africa are responsible for many of its environmental sustainability issues (Higgs and Hill, 2018). Some SMEs have adverse effects on the natural environments, as their business activities and practices pollute and degrade the biodiversity and biosphere (Luken et al., 2019). SMEs' environmental challenges in Africa can be attributed to weak internal environmental protection policies. For example, SMEs' internal environmental policies are weakly adopted to address and reduce solid, gaseous and liquid pollution (Efobi et al., 2018). This suggests that though SMEs in Africa support government efforts towards achieving socio-economic development objectives, they are also responsible for a range of environmental sustainability problems confronting the continent.

The above background has demonstrated that even though some SMEs are currently contributing to the Sustainable Development Goals (SDGs), they also face several challenges. These challenges weaken their efforts to improve their contribution to social, environmental and economic development. Despite their limitations, this chapter seeks to review and synthesise the current literature to understand business sustainability within the context of SMEs in Africa, and furthermore, propose a new research agenda. In so doing, this chapter employs a simplified systematic literature review to synthesise relevant scholarly works. It proposes a research agenda to improve our understanding of business sustainability in African SMEs for students, academics and practitioners.

The chapter is structured as follows: the background of SMEs in Africa is presented by highlighting their characteristics, challenges and contribution towards sustainable development. Next, the relevant literature on business sustainability and SMEs in Africa is explored and reviewed. This is followed by a synthesis of the literature on business sustainability towards developing an Afrocentric research agenda. This agenda highlights key themes that students, educators, researchers and practitioners can discuss and investigate to strengthen African research on SMEs' business sustainability. The chapter concludes by highlighting the significance of SMEs in achieving the SDGs in Africa.

2.   SMEs: characteristics, contributions and challenges

2.1   Characteristics

The review of the relevant literature reveals that Africa lacks a universal definition and classification of SMEs. As shown in Table 10.1, available information indicates that the South African SMEs classification system is the most cited scale for defining SMEs (Abor and Quartey, 2010).

Most SMEs in Africa are geographically dispersed. These businesses are scattered across rural and urban centres, and urban SMEs are generally understood as more organised than rural ones (Kayanula and Quartey, 2000). However, the majority of SMEs in African countries are largely rural and disorganised enterprises operating in diverse sectors and industries. For example, rural SMEs operate in the manufacturing, agriculture and mining industries (UNECA, 2010). Urban SMEs also engage in mining, manufacturing, agricul-

Table 10.1   Definitions of SMEs as provided in the South African National Small Business Act (1996)

Enterprise Size	Number of Employees	Annual Turnover (in South African rand [R])	Gross Assets, Excluding Fixed Property
Medium	Fewer than 100 to 200, depending on industry	Less than R4 million to R50 million, depending on industry	Less than R2 million to R18 million, depending on industry
Small	Fewer than 50	Less than R2 million to R25 million, depending on industry	Less than R2 million to R4.5 million, depending on industry
Very Small	Fewer than 10 to 20, depending on industry	Less than R200 000 to R500 000, depending on industry	Less than R150 000 to R500 000, depending on Industry
Micro	Fewer than 5	Less than R150 000	Less than R100 000

Source: Falkena et al. (2001).

ture and fishing industries, as well as the service sector providing goods and services.

2.2   Contributions

The review of published literature has shown that SMEs in Africa contribute to socio-economic development. In Kenya, SMEs contribute 40% of the country's GDP, over 50% of new jobs and 80% of the workforce (Muriithi, 2017). In Nigeria, SMEs contribute 70% to industrial jobs (Muriithi, 2017). In Ghana, SMEs generate income and employment, mobilise funds for banks, pay taxes, and provide goods and services, thereby stimulating the economy (Abor and Quartey, 2010). In Zambia, SMEs provide social services and produce goods for domestic consumption (Nuwagaba, 2015). Similarly, in Tanzania, SMEs contribute 60% to the country's gross national product and support capital markets with 5 million TZ shillings (Mbuyisa and Leonard, 2016). Also, in South Africa, SMEs account for 91% of local businesses, contribute between 50% and 60% of GDP and provide about 60% employment (Ngek, 2014). Despite their socio-economic significance, Higgs and Hill (2018) reveal that SMEs are now supporting environmental initiatives, especially in South Africa.

2.3   Challenges

The review of existing literature revealed that SMEs in African countries face challenges that impede sustainability practices. In Ghana, SMEs are constrained by a lack of resources because addressing environmental issues is an expensive venture especially for those with moderately less business capital (Adomako et al., 2021). In South Africa, SMEs face ethical challenges, including bribery and corruption scandals (Painter-Morland and Dobie, 2009). Many SMEs in Africa lack business competitiveness and sustainability because of their limited potential for growth and expansion. Likewise, in Ghana, SMEs face certain challenges such as social, cultural, legal and political challenges that influence sustainability, innovation and global expansion (Takyi and Naidoo, 2020). The challenges indicated above partly explain why SMEs are unable to significantly contribute to sustainable development in Africa.

3.   Business and sustainability literature

3.1   Dominant African perspectives on business sustainability

The most common and emerging standpoints on business sustainability in Africa are regulatory, self-preservation and strategic perspectives. The regulatory perspective highlights the need for SMEs to comply with environmental regulations. This view focuses on building strong institutions and regulatory frameworks that promote sustainability (e.g., Muriithi and Louw, 2017). Closely related to the regulatory perspective is the self-preservation perspective. This perspective argues that SMEs can survive and experience long-term growth if they integrate sustainability principles into their business models (e.g., Sejjaaka et al., 2015; Ligthelm, 2010). Lastly, the strategic perspective focuses on strategic thinking. It emphasises that SMEs can achieve a competitive advantage if their strategic analysis, formulation, implementation and evaluation processes integrate sustainability principles (e.g., Nwoba et al., 2021; Valente, 2015). These three perspectives depict present mindsets and philosophies regarding SMEs' business sustainability in Africa.

3.2   Sustainability practices of SMEs in Africa

The synthesis of the literature highlights several sustainability practices among SMEs in Africa. SMEs are becoming increasingly environmentally conscious. For example, some South African SMEs are environmentally conscious about the effects of their operations and are currently participating in waste management initiatives (Higgs and Hill, 2018). Also, in Ghana, Adomako (2020) observes that some SMEs collaborate with stakeholders to address waste management and environmental sustainability issues. Findings from Kenya and Senegal indicate that some SMEs are building capacities to respond to climate risks associated with their businesses (Crick et al., 2018).

SMEs in Africa also have social significance. In South Africa, Masocha (2019) found that SMEs employ people with low education and skills, including women, in the lower spectrums of society. In Nigeria, Apulu and Latham (2011) posit that SMEs enhance people's living standards through local capital formation and increased productivity and capacity. These Nigerian SMEs aim to eradicate poverty and poor healthcare delivery (Amaeshi et al., 2016). In Kenya, most SMEs, through social responsibility, are addressing societal concerns such as HIV/AIDS, health and education (Muthuri and Gilbert, 2011).

Furthermore, SMEs in Africa have economic significance. For instance, in Ghana, SMEs are considered the country's economic engine as their presence stimulates the economy (Abor and Quartey, 2010). Similarly, in South Africa, SMEs are solving economic problems, including unemployment and poverty by creating jobs (Ladzani and Netswera, 2009). South African SMEs' economic participation is vital for achieving local economic development through job creation, poverty eradication and economic empowerment of disadvantaged people (Fiseha and Oyelana, 2015). In other sub-Saharan African countries, SMEs continue to play a significant part in poverty reduction and job creation, especially within impoverished countries, such as Liberia, Mauritania, Gabon, Togo and Benin (Abisuga-Oyekunle et al., 2020).

3.3   Key drivers of sustainability practices in SMEs

According to the literature analysis, sustainability practices in SMEs can be attributed to the following factors. In South Africa, SMEs are driven by government directives to participate in waste management programmes (Higgs and Hill, 2018). They are also driven by environmental regulations and legal compliance to improve their sustainability practices (Godfrey et al., 2017). On the other hand, in Ghana, studies have revealed that SMEs are driven by financial and economic performance to be environmentally sustainable (Amankwah-Amoah et al., 2019; Danso et al., 2019). Further, in Kenya and Senegal, Crick et al. (2018) indicate that SMEs are driven by international concern about climate change and climate risks to adopt and adapt sustainability principles. However, in South Africa, SMEs practise sustainability because of internal control systems put in place by management to support government policies and regulations (Bruwer et al., 2018). Furthermore, in Egypt most SMEs engage in sustainability because of pressures from stakeholders (Aboelmaged,

2018). These drivers explain SMEs attitudes and behaviours towards sustainability in Africa.

4.   An Afrocentric research agenda

An Afrocentric research agenda was developed by reviewing and synthesising relevant literature on SMEs and sustainability in Africa. The literature analysis showed that developing a research agenda for SMEs and sustainability in Africa requires probing the following five themes: (i) institutions; (ii) modern technologies; (iii) business ethics; (iv) internal control systems and managerial behaviours; and (v) indigenous management. These themes are relevant to students, educators, researchers, policymakers and practitioners interested in SMEs and sustainability in Africa. These emerging themes are discussed below and include relevant questions to be addressed in future research.

4.1   Institutions

Institutions emerged as a common theme during the literature analysis. For example, in South Africa, research findings emphasise the need for strong institutions and regulatory environments that support and promote sustainability in SMEs (Masocha and Fatoki, 2018). Similarly, in Ghana, formal and informal institutions have been found to support SMEs' development (Takyi and Naidoo, 2020). Despite the importance of institutions, our understanding of how these institutions shape SMEs' commitment and participation levels in integrating sustainability principles into their business models is limited (Masocha and Fatoki, 2018). An in-depth understanding of how institutions support business sustainability in most SMEs requires relevant stakeholders to find answers to the following questions:

- What is the origin of institutional frames for sustainability in SMEs in Africa?
- Which western-based institutions can help promote sustainability in African SMEs?
- How can institutions shape SMEs' commitment and participation levels in sustainability in Africa?
- Do institutional void and rigidity account for unsustainable business practices in SMEs in Africa?
- Does the promotion of sustainability in SMEs require a new agency in Africa?

4.2   Modern technologies

SMEs in Africa are embracing modern technologies as a tool for business development and competitive advantage. In Ghana, Amankwah-Amoah (2019) posits that SMEs are adopting modern technologies and their associated tools to become competitive. In Nigeria, Eze and colleagues (2018) found that managers of SMEs are developing productive information behaviour by adopting information and communication technology (ICT). Likewise, in South Africa, Mohlameane and Ruxwana (2013) argue that ICT solutions can increase SMEs competitiveness, thus contributing to business sustainability. There is uncertainty, however, about how SMEs adopt, adapt and use modern technologies to address sustainability issues (Mohlameane and Ruxwana, 2013). Explaining this uncertainty requires answers to the following questions:

- What kinds of modern technologies can support SMEs' sustainability in Africa?
- What kinds of internal and external environmental factors can drive a technological revolution in SMEs in Africa?
- How can SMEs in Africa invest in green technology?
- How can SMEs in Africa promote technological revolution?

4.3   Business ethics

Ethical issues in SMEs emerged as a major concern during the analysis of the literature. For instance, in Nigeria, Cant et al. (2013) indicate that one of the biggest threats to SMEs is employee theft, resulting in about 30% of small business failures. Additionally, in Ghana, Sackey et al. (2013) have found that unethical business practices occur when applying for business information and permits, competing for business contracts and funds, and dealing with tax authorities. Similarly, Painter-Morland and Dobie (2009) argue that most South African SMEs face bribery and corruption issues that adversely affect their contributions to sustainability. Contrarily, SMEs managers in Cameroon claim that they are not responsible for the natural environment and other widespread unethical business practices (Demuijnck and Ngnodjom, 2013). Understanding how these ethical and unethical issues affect SMEs' sustainability require answers to these questions:

- What explains unethical behaviours in SMEs across different sectors in Africa?
- What is the ethical orientation of SMEs owners and managers in Africa?
- How do sociocultural environments shape SMEs' business ethics?
- How does ethics shape SMEs' sustainability in different African countries?

4.4 Internal control systems/managerial behaviours

Internal control systems and managerial behaviours were identified as critical factors for SME development. In this light, Bure and Tengeh (2019) argue that internal control systems could serve as preventative measures to stop the failure of SMEs in Zimbabwe. In South Africa, Bruwer et al. (2018) also found that good internal control systems such as internal control activities and managerial conduct affect SME sustainability. Moreover, in Nigeria, Adepoju (2017) opines that most SMEs use segregation of duty and adherence to processes, policies and procedures to improve internal control practices. In Ghana, Tetteh et al. (2020) also showed that internal control systems can improve the performance and profitability of listed SMEs. Addressing the following questions can help improve our understanding of how internal control systems and managerial behaviours affect sustainability of SMEs in Africa:

- What kinds of internal control systems are used by SMEs in Africa?
- How can internal control systems influence SMEs' sustainability in different sectors in Africa?
- What kind of managerial competencies exist in SMEs in Africa?
- Do managerial competencies support the internal control systems of SMEs in Africa?
- How do managerial competencies influence the sustainability of SMEs in different sectors in Africa?

4.5 Indigenous management

Scholars in Africa contend that management issues need to be addressed using indigenous African management philosophies rooted in the African culture, value systems and beliefs (Inyang, 2008). Jackson et al. (2008) found that both inter-continental and intra-country cultural differences strongly influence the success of local SMEs in Kenya. In Ghana, Amoako and Matlay (2015) found that indigenous norms shape trust and relationship building among export-based SMEs, while Adeodu et al. (2015) highlight the development of indigenous technology through engineering-based SMEs to serve as capacity building for sustainable development. In South Africa, Witbooi and Ukpere (2011) opine that historically racial policies on poverty, employment and income levels affect indigenous female entrepreneurs' ability to access finance. Exploring the link between indigenous management and sustainability of SMEs in Africa require answers to the following questions:

- How does western knowledge differ from indigenous knowledge in the African business context?
- How does western knowledge support SMEs' sustainability in Africa?

- What types of indigenous knowledge support SMEs' sustainability and which ones do not?
- How does indigenous knowledge shape management practices in African SMEs?
- In what ways do western management practices influence SMEs' sustainability in Africa?

5. Proposed future methodological approaches

Future research examining the thematic areas stated above can apply different methodological approaches, as shown in Table 10.2. With regards to institutions, those supporting SMEs in Africa have been largely assessed using quantitative approaches. Quantitative approaches such as surveys, observations and case studies were useful for examining formal institutions, whereas informal institutions are examined using qualitative methodology. For example, survey designs can provide in-depth insights into how institutions foster business sustainability for SMEs in Africa (Masocha and Fatoki, 2018). The application of modern technologies to solve SMEs' sustainability issues could be achieved using action research approaches. Action research is defined as a rigorous and disciplined approach to generating knowledge and theory in social science aimed at driving social action in the service of addressing specific and real challenges (Hind et al., 2013). Business ethics have been explored using qualitative and conceptual approaches. These approaches were used in the initial exploration of some factors that influence SMEs' business ethics and sustainability (Demuijnck and Ngnodjom, 2013). Future testing and validation of these factors might require quantitative tools.

Research examining internal control systems and managerial behaviours usually use mixed-method approaches (Bruwer et al., 2018). In addition, interviews and surveys are used to gather empirical data on perceptions of internal control activities and financial positions. However, a deeper understanding of internal control activities and managerial behaviours in relation to business sustainability requires qualitative approaches. Qualitative approaches can supplement the present understanding of internal control activities and management behaviours from the positivist perspective. Comprehension of the behavioural tendencies of managers and perceived internal control activities of SMEs is also possible through interpretivist traditions. Indigenous management practices are primarily informal social processes (Inyang, 2008); hence, an in-depth understanding of how informal processes support SMEs' sustainability could be possible through qualitative approaches. Informal social

Table 10.2 Proposed future methodological approaches

Themes	Method	Data Type/Source	Approach
Institutions: • Formal Institutions • Informal Institutions	Mixed Method Quantitative Qualitative	Survey In-depth Interview	Positivist Interpretive
Modern Technologies	Action Research	In-depth Interview	Interpretive
Business Ethics	Multiple Method Qualitative Conceptual Quantitative	In-depth Interview Literature Review Survey	Interpretive Systematic Review Positivist
Internal Control Systems & Managerial Behaviours	Mixed Method	In-depth Interview Survey	Interpretive Positivist
Indigenous Management	Qualitative	In-depth Interview	Interpretive

Source: Author.

interactions and processes provide opportunities for idea sharing, resulting in creative solutions for addressing SMEs' sustainability.

6. Concluding remarks

This chapter examines the current understanding of SMEs' sustainability in Africa to develop and propose an Afrocentric research agenda. Thus, a synthesis of the relevant literature on SMEs' business sustainability has demonstrated that SMEs are integrating sustainability principles into their business models despite their unique challenges. This chapter demonstrates that though SMEs are making efforts to integrate sustainability principles into their business practices, very little attention has been given to business sustainability in the SME literature in Africa. More precisely, the emerging themes identified have received little conceptual, theoretical and empirical assessment in African SMEs. This growing gap requires students, educators, researchers and policymakers to investigate the identified themes to improve the sustainability of African businesses in general and SMEs in particular. Nevertheless, given the contributions and roles of SMEs in the sustainable development of African countries, as the literature demonstrates, it is evident that more needs to be done in promoting scholarly works on business sustainability and integrating sustainability principles into SMEs' operations and strategies.

References

Abisuga-Oyekunle, O. A., Patra, S. K. and Muchie, M. (2020), "SMEs in sustainable development: Their role in poverty reduction and employment generation in sub-Saharan Africa", *African Journal of Science, Technology, Innovation and Development*, Vol. 12 No. 4, pp. 405-419.

Aboelmaged, M. (2018), "The drivers of sustainable manufacturing practices in Egyptian SMEs and their impact on competitive capabilities: A PLS-SEM model", *Journal of Cleaner Production*, Vol. 175, pp. 207-221. https:// doi .org/ 10 .1016/ j .jclepro.2017.12.053

Abor, J. and Quartey, P. (2010), "Issues in SME development in Ghana and South Africa", *International Research Journal of Finance and Economics*, Vol. 39 No. 6, pp. 215-228.

Adeodu, A., Daniyan, I., Omohimoria, C. and Afolabi, S. (2015), "Development of indigenous engineering and technology in Nigeria for sustainable development through the promotion of SMEs (case of design of manually operated paper recycling plant)", *International Journal of Science, Technology and Society*, Vol. 3 No. 4, pp. 124-131. https://doi.org/10.11648/j.ijsts.20150304.15

Adepoju, A. O. (2017), "Strategies for improving internal control in small and medium enterprises in Nigeria", Doctoral Dissertation, University of Walden, United States.

Adomako, S. (2020), "Environmental collaboration, sustainable innovation, and small and medium-sized enterprise growth in sub-Saharan Africa: Evidence from Ghana", *Sustainable Development*, Vol. 28 No. 6, pp. 1609-1619. https://doi.org/10.1002/sd .2109

Adomako, S., Ning, E. and Adu-Ameyaw, E. (2021), "Proactive environmental strategy and firm performance at the bottom of the pyramid", *Business Strategy and the Environment*, Vol. 30 No. 1, pp. 422-431. https://doi.org/10.1002/bse.2629

Amaeshi, K., Adegbite, E., Ogbechie, C., Idemudia, U., Kan, K. A. S., Issa, M. and Anakwue, O. I. (2016), "Corporate social responsibility in SMEs: A shift from philanthropy to institutional works?", *Journal of Business Ethics*, Vol. 138 No. 2, pp. 385-400. https://doi.org/10.1007/s10551-015-2633-1

Amankwah-Amoah, J. (2019), "Technological revolution, sustainability, and development in Africa: Overview, emerging issues, and challenges", *Sustainable Development*, Vol. 27 No. 5, pp. 910-922. https://doi.org/10.1002/sd.1950

Amankwah-Amoah, J., Danso, A. and Adomako, S. (2019), "Entrepreneurial orientation, environmental sustainability and new venture performance: Does stakeholder integration matter?", *Business Strategy and the Environment*, Vol. 28 No. 1, pp. 79-87. https://doi.org/10.1002/bse.2191

Amoako, I. O. and Matlay, H. (2015), "Norms and trust-shaping relationships among food-exporting SMEs in Ghana", *The International Journal of Entrepreneurship and Innovation*, Vol. 16 No. 2, pp. 123-134. https://doi.org/10.5367/ijei.2015.0182

Apulu, I. and Latham, A. (2011), "Drivers for information and communication technology adoption: A case study of Nigerian small and medium sized enterprises", *International Journal of Business and Management*, Vol. 6 No. 5, pp. 51-60. https://doi.org/10.5539/ijbm.v6n5p51

Bruwer, J.-P., Coetzee, P. and Meiring, J. (2018), "Can internal control activities and managerial conduct influence business sustainability? A South African SMME perspective", *Journal of Small Business and Enterprise Development*, Vol. 25 No. 5, pp. 710-729. https://doi.org/10.1108/JSBED-11-2016-0188

Bure, M. and Tengeh, R. K. (2019), "Implementation of internal controls and the sustainability of SMEs in Harare in Zimbabwe", *Entrepreneurship and Sustainability Issues*, Vol. 7 No. 1, pp. 201-218. https://doi.org/10.9770/jesi.2019.7.1(16)

Cant, M. C., Wiid, J. A. and Kallier, S. M. (2013), "Small business owners' perceptions of moral behaviour and employee theft in the small business sector of Nigeria", *Gender and Behaviour*, Vol. 11 No. 2, pp. 5775-5787.

Crick, F., Eskander, S. M., Fankhauser, S. and Diop, M. (2018), "How do African SMEs respond to climate risks? Evidence from Kenya and Senegal", *World Development*, Vol. 108, pp. 157-168. https://doi.org/10.1016/j.worlddev.2018.03.015

Danso, A., Adomako, S., Amankwah-Amoah, J., Owusu-Agyei, S. and Konadu, R. (2019), "Environmental sustainability orientation, competitive strategy and financial performance", *Business Strategy and the Environment*, Vol. 28 No. 5, pp. 885-895. https://doi.org/10.1002/bse.2291

Demuijnck, G. and Ngnodjom, H. (2013), "Responsibility and informal CSR in formal Cameroonian SMEs", *Journal of Business Ethics*, Vol. 112 No. 4, pp. 653-665. https://doi.org/10.1007/s10551-012-1564-3

Efobi, U., Belmondo, T., Orkoh, E., Atata, S. N., Akinyemi, O. and Beecroft, I. (2018), "Environmental pollution policy of small businesses in Nigeria and Ghana: Extent and impact", *Environmental Science and Pollution Research*, Vol. 26 No. 3, pp. 2882-2897. https://doi.org/10.1007/s11356-018-3817-x

Eze, S. C., Olatunji, S., Chinedu-Eze, V. C. and Bello, A. O. (2018), "Key success factors influencing SME managers' information behaviour on emerging ICT (EICT) adoption decision-making in UK SMEs", *The Bottom Line*, Vol. 31 No. 3/4, pp. 250-275. https://doi.org/10.1108/BL-02-2018-0008

Falkena, H., Abedian, I., Blottniz, M., Coovadia, C., Davel. G., Madungandaba, J., Masilela, E. and Rees, S. (2001), "SMEs' access to finance in South Africa: A supply-side regulatory review", *The Task Group of the Policy Board for Financial Services and Regulation* (pp. 1-316), Pretoria, South Africa.

Fiseha, G. G. and Oyelana, A. A. (2015), "An assessment of the roles of small and medium enterprises (SMEs) in the local economic development (LED) in South Africa", *Journal of Economics*, Vol. 6 No. 3, pp. 280-290. https://doi.org/10.1080/09765239.2015.11917617

Godfrey, L., Muswema, A., Strydom, W., Mamafa, T. and Mapako, M. (2017), "Co-operatives as a development mechanism to support job creation and sustainable waste management in South Africa", *Sustainability Science*, Vol. 12 No. 5, 799-812.

Higgs, C. J. and Hill, T. (2018), "The role that small and medium-sized enterprises play in sustainable development and the green economy in the waste sector, South Africa", *Business Strategy and Development*, Vol. 2 No. 1, pp. 25-31. https://doi.org/10.1002/bsd2.39

Hind, P., Smit, A. and Page, N. (2013), "Enabling sustainability through an action research process of organisational development", *Journal of Corporate Citizenship*, Vol. 49, pp. 137-161.

Inyang, B. J. (2008), "The challenges of evolving and developing management indigenous theories and practices in Africa", *International Journal of Business and Management*, Vol. 3 No. 12, pp. 122-132. https://doi.org/10.5539/ijbm.v3n12p122

Jackson, T., Amaeshi, K. and Yavuz, S. (2008), "Untangling African indigenous management: Multiple influences on the success of SMEs in Kenya", *Journal of World Business*, Vol. 43 No. 4, pp. 400-416. https://doi.org/10.1016/j.jwb.2008.03.002

Kamunge, M. S., Njeru, A. and Tirimba, O. I. (2014), "Factors affecting the performance of small and micro enterprises in Limuru Town Market of Kiambu County, Kenya", *International Journal of Scientific and Research Publications*, Vol. 4 No. 12, pp. 1-20.

Kayanula, D. and Quartey, P. (2000), "The policy environment for promoting small and medium-sized enterprises in Ghana and Malawi", Finance and Development Research Programme, Working Paper Series, Paper No 15, IDPM, University of Manchester.

Ladzani, W. and Netswera, G. (2009), "Support for rural small businesses in Limpopo Province, South Africa", *Development Southern Africa*, Vol. 26 No. 2, pp. 225-239. https://doi.org/10.1080/03768350902899512

Ligthelm, A. A. (2010), "Entrepreneurship and small business sustainability", *Southern African Business Review*, Vol. 14 No. 3, pp. 131-153.

Luken, R. A., Clarence-Smith, E., Langlois, L. and Jung, I. (2019), "Drivers, barriers, and enablers for the greening industry in sub-Saharan African countries", *Development Southern Africa*, Vol. 36 No. 5, pp. 570-584. https://doi.org/10.1080/0376835X.2018.1503944

Masocha, R. (2019), "Social sustainability practices on small businesses in developing economies: A case of South Africa", *Sustainability*, Vol. 11 No. 12, pp. 1-13. https://doi.org/10.3390/su11123257

Masocha, R. and Fatoki, O. (2018), "The impact of coercive pressures on sustainability practices of small businesses in South Africa", *Sustainability*, Vol. 10 No. 9, pp. 1-14. Doi:10.3390/su10093032 https://doi.org/10.3390/su10093032

Mbuyisa, B. and Leonard, A. (2016), "The role of ICT use in SMEs towards poverty reduction: A systematic literature review", *Journal of International Development*, pp. 1-39. doi:10.1002/jid

Mohlameane, M. J. and Ruxwana, N. L. (2013), "The potential of cloud computing as an alternative technology for SMEs in South Africa", *Journal of Economics, Business and Management*, Vol. 1 No. 4, pp. 396-400. https://doi.org/10.7763/JOEBM.2013.V1.85

Muriithi, S. (2017), "African small and medium enterprises (SMEs) contributions, challenges and solutions", *European Journal of Research and Reflection in Management Science*, Vol. 5 No. 1, pp. 1-14.

Muriithi, S. M. and Louw, L. (2017), "The Kenyan banking industry: Challenges and sustainability", Ahmed, A. (Ed.), in *Managing Knowledge and Innovation for Business Sustainability in Africa*, Palgrave Studies of Sustainable Business in Africa. Palgrave Macmillan, Cham. https://doi.org/10.1007/978-3-319-41090-6_11

Muthuri, J. N. and Gilbert, V. (2011), "An institutional analysis of corporate social responsibility in Kenya", *Journal of Business Ethics*, Vol. 98 No. 3, pp. 467-483. https://doi.org/10.1007/s10551-010-0588-9

Ngek, N. B. (2014), "Determining high quality SMEs that significantly contribute to SME growth: Regional evidence From South Africa", *Problems and Perspectives in Management*, Vol. 12 No. 4, pp. 1-13.

Nuwagaba, A. (2015), "Enterprises (SMEs) in Zambia", *International Journal of Economics, Finance and Management*, Vol. 4 No. 4, pp. 146-153.

Nwoba, A. C., Boso, N. and Robson, M. J. (2021), "Corporate sustainability strategies in institutional adversity: Antecedent, outcome, and contingency effects", *Business Strategy and the Environment*, Vol. 30 No. 2, pp. 787-807. https://doi.org/10.1002/bse.2654

Oguntoye, O. and Evans, S. (2017), "Framing manufacturing development in Africa and the influence of industrial sustainability", *Procedia Manufacturing*, Vol. 8, pp. 75-80. https://doi.org/10.1016/j.promfg.2017.02.009

Painter-Morland, M. and Dobie, K. (2009), "Ethics and sustainability within SMEs in sub-Saharan Africa: Enabling, constraining and contaminating relationships", *African Journal of Business Ethics*, Vol. 4 No. 2, pp. 7-19. https://doi.org/10.15249/4-2-66

Sackey, J., Fältholm, Y. and Ylinenpää, H. (2013), "Working with or against the system: Ethical dilemmas for entrepreneurship in Ghana", *Journal of Developmental Entrepreneurship*, Vol. 18 No. 1, pp. 1-18. https://doi.org/10.1142/S1084946713500052

Sampath, P. (2016), "Sustainable industrialization in Africa: Toward a new development agenda", Oyelaran-Oyeyinka, B. and Sampath, P. (Eds), in *Sustainable Industrialization in Africa: Towards a New Development Agenda*, Palgrave Macmillan, London, pp. 1-19.

Schmidheiny, S. (1992), *Changing Course: A Global Business Perspective on Development and the Environment*, MIT Press, Cambridge, MA.

Sejjaaka, S., Mindra, R. and Nsereko, I. (2015), "Leadership traits and business sustainability in Ugandan SMEs: A qualitative analysis", *International Journal of Management Science and Business Administration*, Vol. 1 No. 6, pp. 42-57. https://doi.org/10.18775/ijmsba.1849-5664-5419.2014.16.1004

Takyi, L. and Naidoo, V. (2020), "Innovation and business sustainability among SMEs in Africa: The role of the institutions", Gao, Y., Tsai, S., Du, X., and Xin, C. (Eds), in *Sustainability in the Entrepreneurial Ecosystem: Operating Mechanisms and Enterprise Growth*, IGI Global, Pennsylvania, United States, pp. 50-74.

Tetteh, L. A., Kwarteng, A., Aveh, F. K., Dadzie, S. A. and Asante-Darko, D. (2020), "The impact of internal control systems on corporate performance among listed firms in Ghana: The moderating role of information technology", *Journal of African Business*, pp. 1-22. https://doi.org/10.1080/15228916.2020.1826851

United Nations Economic Commission for Africa (UNECA) (2010), "Progress towards sustainable development report in Africa", UNECA, Addis Ababa.

Valente, M. (2015), "Business sustainability embeddedness as a strategic imperative: A process framework", *Business & Society*, Vol. 54 No. 1, pp. 126-142. https://doi.org/10.1177/0007650312443199

Witbooi, M. and Ukpere, W. (2011), "Indigenous female entrepreneurship: Analytical study on access to finance for women entrepreneurs in South Africa", *African Journal of Business Management*, Vol. 5 No. 14, pp. 5646-5657.

11   Sustainability management tools: value of reporting and assurance

Kari Solomon, Sally V. Russell and Effie Papargyropoulou

1.   Introduction and research background

The Triple Bottom Line (TBL) is an early adopted and frequently used tool to examine the relationship between the environmental, social, and economic impacts of a company (Elkington, 1994). However, through its use, the focus of the tool has moved from examining impacts towards prioritising the profitability and financial performance of the firm (Elkington, 2018). There is concern in the literature that the TBL approach to sustainability management will continue to be diluted from its original aim without consistency across reporting standards (Christofi, Christofi and Sisaye, 2012). Many organisations are adopting sustainability management tools as a means of expanding business strategy beyond financial results. Sustainability management tools can support the functional improvement of an organisation's sustainability commitments through implementation of performance measurement practices (Klovienė and Speziale, 2015). To best understand the impact of an organisation's sustainability practices beyond their financial performance, performance management and assurance practices help establish the link between an organisation's sustainability action and the stakeholders affected by the result of that action.

An organisation can choose from a number of different sustainability management tools. Most standards are voluntary, and mandatory reporting is largely based on financial reporting processes. While having a varied set of tools to pick from may appear advantageous, one of the key issues is that variety creates an overwhelming number of options for organisations. Overwhelming options discourage organisations from making meaningful contributions to sustainable development. As well as having a variety of tools, the lack of global standards or processes for reporting creates a mixed use of sustainability management tools and outcomes that are not comparable and remain difficult to test and valuate (Machado, Dias and Fonseca, 2021). Research focused on understanding how these tools are used often leads to seeking a standard approach to evaluate sustainability reporting. There is a growing demand for establishing the credibility of sustainability, where corporate governance and organisational culture inform how sustainable development actions and investments may affect societal behaviours (Klovienė and Speziale, 2015).

Organisations reporting their investment activities in the context of social impact may contribute to long-term changes in societal behaviour that align with sustainable development principles (Camilleri, 2018). Sustainability reporting provides information beyond the typical financial reporting intended for shareholders and broader sets of stakeholders (Klovienė and Speziale, 2015). One form of evaluation comes in participating in an assurance or audit process, which verifies and evaluates conformance and performance. The purpose and outputs of reporting practices continues to evolve. It is predicted that assurance will become a more complex process in evaluating the relevance of report content to stakeholder expectations (Camilleri, 2018).

Performance and assurance practices, however, are still relatively new and currently not closely aligned with principles of existing sustainability management tools that would ensure the necessary level of evaluation (Boiral, Heras-Saizarbitoria and Brotherton, 2019). Organisations developing comprehensive sustainability strategies that include stakeholder requirements, can seek the assurance of sustainability reporting as part of their commitment to transparency, quality, and accessibility of information about the impact of their sustainability practices. However, this is as yet not fully developed. In the following sections we will review the most commonly used sustainability management tools and explore concepts of sustainability report assurance and suggest important avenues for future research.

2. Sustainability management tools

Sustainability management tools are often presented in the form of reporting standards or frameworks to help organisations communicate their contribution to sustainable development. As more organisations provide sustainability reports, the quality and accuracy of reporting becomes increasingly important. Some larger organisations engage assurance providers like Deloitte or PricewaterhouseCoopers in the sustainability reporting process to provide a quality assurance and checking mechanism for content published in the sustainability reports.

However, assurance providers may be less likely to take sustainable development principles seriously in the reporting process. Assurance providers use a rule-based approach relating to financial standards which has not yet adapted to the evaluation of information on sustainability issues (Boiral, Heras-Saizarbitoria and Brotherton, 2019). There is an opportunity to explore how assurance providers learn and integrate principles from various sustainability management tools, like the Task-force for Climate-related Financial Disclosure (TCFD), Global Reporting Initiative (GRI), or Climate Disclosure Standards Board (CDSB), to build criteria that evaluates conformance and quality of content in an organisation's sustainability report.

The CDSB and the GRI are two of the most agile frameworks with reporting principles built in line with the Sustainable Development Goals (SDGs). Auditing integrated reporting standards for organisations with diverse stakeholder groups adds validity and accountability. It achieves this by requiring openness in strategic financial positioning to generate shared value for stakeholders (Camilleri, 2018). A key cause for concern in the availability of so many reporting standards is the extent to which they are congruent with sustainable development principles. New standards are constantly in development, which provides a capacity issue for organisations that use multiple sustainability management tools to communicate strategic decisions for sustainability.

Effective use of sustainability management tools results in reliable governance and quality reporting of sustainability practices and outcomes. An organisation with a reliable governance structure can improve their reporting using an integrated reporting standard, which combines financial and sustainability reporting practices into an aligned report (Hamad, Draz and Lai, 2020). Organisations that integrate sustainable development principles into their corporate governance structure better leverage sustainability management tools to report key performance indicators of social and environmental impacts.

Communicating sustainability issues does not necessarily mean publishing an accurate report. An organisation's information reported in compliance to a sustainability standard, like GRI or TCFD, does not equate to the information itself being valuable, accurate, or useful to stakeholders (Haller, van Staden and Landis, 2018). As the emphasis shifts from compliance to performance, a new field of study emerges: sustainability reporting assurance, in which the transparency, accuracy, and relevance of reporting are validated against a standard and used to compare the performance of sustainability initiatives across organisations.

The standard for sustainability assurance is yet to be established. Other stakeholder groups, such as employees, communities, or investors, exert more pressure on organisations to improve the credibility of sustainability reporting and strategy than regulatory agencies (Aureli et al., 2020). In the following sections, we review some of the most commonly used sustainability management tools, outlining each tool's structure and application, and identifying the most pressing research questions. Additionally, we provide insight and critique of how some management tools are being consolidated for future strategy and reporting best practice.

2.1   Global Reporting Initiative

The GRI sustainability reporting standards are designed for organisations to report the impact of their activities on the economy, environment, and society (GRI, 2020, p. 3). The GRI seeks to establish a reporting framework usable by organisations of any sector, size, or location. The GRI is one of many reporting standards used to homogenise sustainability reporting, and is most closely aligned to TBL. The GRI established value and transparency in reporting and assurance, through ensuring a complete materiality assessment process is in place (Machado, Dias and Fonseca, 2021). In the context of sustainability, the GRI defines materiality as "the principle that determines which relevant topics are sufficiently important that it is essential to report on them" (GRI, 2020).

The materiality assessment considers an organisation's business strategy and stakeholder expectations to prioritise sustainability issues for action. Reporting to GRI standards supports organisations to integrate sustainable development principles into their business practices, and select impactful social, environmental, and economic performance metrics. The materiality process outlined in the GRI provides a standard management approach for organisations to integrate environmental, social, and economic considerations into their business strategy (Hayatun, Burhan and Rahmanti, 2012). The engagement of stakeholders in the materiality review process ensures that an organisation is aware of the implications of its sustainability practices.

While the GRI is a useful tool for assessing and reporting TBL-based sustainability performance, its application is limited by organisations that use it as a strategy and communication platform to promote the value of sustainability practices. For example, it is still unclear whether the use of sustainability reporting leads to sustainability action, that in turn has positive impact on organisations' stakeholders (Haller, van Staden and Landis, 2018). The GRI as a sustainability management tool provides a structured approach to building strategy and basic performance indicators, yet lacks a clear path to assessing the impact of an organisation's sustainability practices on stakeholders or broader society. Thus, there is a growing need for more research on how GRI reporting organisations estimate and trace the impact of corporate operations on social and environmental aspects external to the business.

2.2 United Nations Sustainable Development Goals

The United Nations (UN) SDGs are one of the most commonly adapted sustainability management tools (GRI, United Nations Global Compact and World Business Council for Sustainable Development, 2016). The SDGs are the outcome of the UN 2030 Agenda established at the 2015 UN General Assembly, focusing on global goals and initiatives for sustainable development (Bergman, Bergman and Berger, 2017). Organisations are looking to public policy and third-party organisations to define sustainability management tools like the SDGs and to aid in developing impactful sustainability strategies.

The SDGs offer a standard for measuring sustainability performance and impact, rather than providing a reporting structure. A pitfall is the lack of a unifying, visionary goal for all committed organisations to align their strategy (Costanza et al., 2016). The SDGs are aligned to the GRI and TCFD reporting standards as made evident in the SDG Compass and SDG Report Integration tools provided by the GRI (GRI, United Nations Global Compact and World Business Council for Sustainable Development, 2016; GRI, 2020). This linkage is additionally framed by the TBL approach. While the SDGs do not provide a reporting process, they are used frequently by organisations to frame objectives of sustainability initiatives within sustainability reports.

Many sustainability management tools are evolving to integrate the SDGs and to inform how impacts and outcomes of organisations are categorised and communicated to stakeholders. The complexity and breadth of the SDGs is a benefit in that they are comprehensive in their coverage of key issues. However, this also creates a barrier for organisations seeking to measure their progress. It is more common for organisations to select a subset of goals and prioritise measuring efforts to attain a more focused set of outcomes (Ike et al., 2019). The SDGs provide a broad array of goals and targets for any organisations to use, but they lack an aligning metric to measure progress towards accomplishing a global goal (Costanza et al., 2016). The continued variation of use and adoption of the SDGs by organisations further dilutes the potential impact they could have on meaningful outcomes of sustainability management in organisations. Future research that examines the effect of such complexity and the level of application for organisations would be of benefit.

2.3   United Nations Global Compact

The United Nations Global Compact (UNGC) was launched in 2000 and is a "voluntary initiative based on CEO commitments to implement sustainability principles and to take steps to support UN goals" (United Nations, 2021, p. 1). The UNGC is a collection of ten principles focused on human and labour rights, environmental preservation, and anti-corruption, that firms can sign up to follow on a voluntary basis. The UNGC differs from the SDGs in that it focuses specifically on corporate sustainability as a mechanism for achieving the SDGs, and encourages businesses to show leadership in advancing the SDGs.

There is evidence that firms adopting the UNGC framework are motivated by ethical and environmental concerns. Firms adopting the UNGC report positive outcomes in terms of their corporate image, network opportunities, sales growth, and profitability benefits (Cetindamar and Husoy, 2007; Orzes et al., 2020). While it is clear that adopting the UNGC principles benefits firms, it is less clear whether the implications for and outcomes on human and labour rights, environmental preservation, and anti-corruption are as significant. More research is needed to identify if and how the UNGC contributes to meaningful business sustainability results beyond financial goals.

2.4   Task-Force for Climate-Related Financial Disclosure

The Financial Standards Board (FSB) established the TCFD in 2015, in response to demands from the financial sector for consistent reporting on climate-related issues. The TCFD offers a framework of recommended guidelines on how to inform shareholders on climate-related financial risks (TCFD, 2017). TCFD focuses on the impact of sustainability practices on financial performance, to demonstrate value creation and risk mitigation to investors and shareholders.

The key reporting areas of TCFD include: governance, strategy, risk, and performance measures, which are similar to the CDSB recommendations. The TCFD provides a mechanism for organisations to report financial risks and opportunities in response to climate change (Eccles, 2018). The TCFD is a key resource for assurance services because it leverages a sustainability reporting framework with the emphasis on demonstrating the financial performance of an organisation's sustainability practices, which may be easier to trace and verify through a formal auditing process.

As more countries and nations adopt mandatory reporting rules for corporate organisations, the TCFD is a framework that prepares organisations for the transition to integrated reporting of impactful sustainability actions (Eccles, 2018). Research in the area of integrated reporting practices shows the trend of sustainability reports published separately from corporate, annual, or finance reports has reduced, and sustainability report content is being included in comprehensive organisation reporting as non-financial disclosure. Pressures from community and public-interest groups on organisations to provide integrated reports results in organisations discussing the links and impacts of their financial and operational strategies on the environment and society (Camilleri, 2018).

2.5   Climate Disclosure Standards Board

Investor demand for corporations to understand the links and repercussions of their actions on environmental and social issues is encouraging organisations to use sustainability management tools. Similar to the TCFD, the Climate Disclosure Standards Board (CDSB) Framework was catalysed by demand for standard reporting practices inclusive of environmental impact and risk management for natural capital (CDSB, 2019). The CDSB Framework offers organisations a reporting standard which supports "decision-useful environmental information" for shareholders and stakeholders (CDSB, 2019, p. 1). Reporting standards are regularly adopted by organisations, as they result in a tangible report to structure discussion on the performance and impact of prioritised sustainability practices in their corporate culture and industry.

The CDSB Framework aims to standardise how organisations include environmental risk-related information into corporate annual and financial reporting processes (CDSB, 2019). Standardising risk-based practices serves the goals of integrated reporting by supporting organisations with an aligned sustainability business strategy. Key reporting areas in the framework include organisational policy, performance, report format, and conformance to the CDSB Framework. Organisations voluntarily using reporting standards like GRI, TCFD, and CDSB may be better positioned to meet future mandatory reporting requirements, as well as prepare for changes in auditing and assurance standards to evaluate the broader performance impact of their sustainability actions.

2.6 Value Reporting Foundation (formerly Sustainability Accounting Standards Board)

The Sustainability Accounting Standards Board (SASB) is a non-profit organisation that was founded in 2011 and published the Conceptual Framework of the Standards in 2017. SASB published approximately 70 industry guidelines in 2018 for reporting on financially significant sustainability issues (Sustainability Accounting Standards Board, 2017). The SASB framework's goal was to provide organisations with reporting tools to help them communicate their value and performance to investors (Sustainability Accounting Standards Board, 2017). Since its inception, the SASB framework has evolved to incorporate the integrated reporting approach, and was re-branded as the Value Reporting Foundation (VRF) in 2019.

The VRF resource distinguishes itself from other sustainability management tools in terms of its user and consumer audiences, with precise protocols for publicly traded organisations to disclose material issues to current and potential investors. The reporting guidelines are developed based on the Securities and Exchange Commission's (SEC) concept of materiality, resulting in fewer sustainability indicators than the GRI or SDGs (Grewal and Serafeim, 2020, p. 95). A more focused standard of performance measures makes the VRF tool more approachable for its intended audience, corporate organisations, increasing adoption and improving performance comparability across industries.

In 2020, the International Financial Reporting Standards (IFRS) Foundation announced a consolidation strategy to assimilate the VRF standards into an aligned, international sustainability financial disclosure standard with the CDSB reporting framework (IFRS, 2021). The IFRS is working to consolidate resources and standards in an effort to engage more organisations with simplified, accessible sustainability management tools that meet demands of investors, shareholders, and customers. The continued consolidation of sustainability management tools may pave the way for new research into concept alignment across financially motivated impact-oriented management and reporting tools.

2.7 Benefit Impact Assessment

B Lab, a non-profit organisation, launched the Benefit Impact Assessment (BIA) in 2007. BIA is a rating and measurement system that evaluates the social and environmental impacts of a company's operations, leading to B Corp Certification for those companies who score well enough on the assessment (Nigri, Michelini and Grieco, 2017). Benefit Corporations, a legal organisational structure, broaden board members' responsibilities by incorporating environmental and social concerns into shareholders' financial interests, but organisations do not need to be structured as a Benefit Corporation to make use of the BIA tool. Such an organisational structure provides executive management legal protection, allowing them to pursue a mission and examine social and environmental impacts of their organisation's operations (Nigri, Michelini and Grieco, 2017). Making use of the BIA contributes to the ethical and social governance standard of an organisation in pursuit of its corporate sustainability strategies. The BIA also provides guidance to an organisation's executive leadership and a reporting standard that is integrated into an organisation's existing performance management system (Nigri and Baldo, 2018).

Organisations can leverage the B Corp Certification process in one of two ways. First, they can fortify or improve the brand management of a company. Second, they can legitimise the outcomes of operations that have social and environmental impact on potential shareholders and investors. The B Corp Certification programme is intended to assist businesses in finding more effective ways to incorporate social ideals into their operations, and more understanding of what corporations do after they have been certified, particularly in terms of stakeholder involvement and corporate governance, is needed (Villela, Bulgacov and Morgan, 2021). The BIA is a sustainability management tool that is similar to the GRI in that it provides a set of guidelines on how to design sustainability strategy and define relevant measures for internal performance, with the addition of an impact evaluation tool to quantify the progress and outputs of sustainability practices.

Organisations look for sustainability management tools which provide a basis for evaluation and comparison of sustainability performance. A report or other type of public reporting confirming an organisation's commitment to and performance of sustainability practices, as well as the impact of activities, is a crucial consequence of all sustainability management tools. More research into the performance and impact of such organisations is required to provide a better understanding of how progress towards sustainable development is measured and tracked. To date, it is unclear how organisations can differentiate their performance and value creation by certifying or restructuring as social enterprises or benefit corporations.

Box 11.1 Case in brief: large corporations implementing sustainability management tools

In sustainability business management research, the key constructs used are institutional, legitimacy, and stakeholder theories (Tavares and Dias, 2018). The combination of these theories provide understanding of organisational decision-making for sustainability strategy and performance. Some organisations are driven by financial risks, some by shareholder demand, and others by regulatory compliance. The use of any sustainability management tool should result in mitigation of risk and improved understanding of environmental and social impacts.

An organisation's decision-making process is key to effectively implement sustainability strategy, and it requires integration of environmental, social, and economic aspects into the core business function (Chofreh and Goni, 2017). Organisations contextualise sustainability management tools as either standards or frameworks. A standard establishes specific metrics and content to be reported, and frameworks provide reporting guidance rather than specific actions or steps (GRI, 2022).

We present the case studies of Marks & Spencer (M&S) and Target Corporation (Target) to illustrate how organisations use sustainability management tools, reporting standards and reporting assurance, to report and manage their sustainability strategy and performance. Table 11.1 briefly outlines which sustainability management tools are used by M&S and Target.

Table 11.1 Sustainability management tools used by M&S and Target

Sustainability Management Tool	Marks & Spencer	Target Corporation
Frameworks	Sustainable Development Goals	Sustainable Development Goals Environmental Social Governance United Nations Guiding Principles Reporting Framework
Reporting Standards	Global Reporting Initiative Value Reporting Foundation United Nations Global Compact [Content Index]	Global Reporting Initiative Value Reporting Foundation Task-force of Climate-related Financial Disclosure
Reporting Assurance	International Standards on Assurance Engagement (ISAE) 3000 via third party	No Assurance Statement Provided

Source: Marks and Spencer (2021); The Target Corporation (2021).

M&S and Target are using a blended approach to effectively manage their sustainability strategy and performance, typically researched under the legitimacy and stakeholder theories. Blending these particular sustainability management tools enables these organisations to align financially motivated and impact-motivated success factors to meet expectations of key stakeholders and report performance. M&S adds credibility in their report by using third-party assurance services, summarised in an assurance statement at the end of their report. The purpose of the restatement is to satisfy investors and shareholder expectations in M&S's ability to mitigate risk and reduce negative impacts (Marks and Spencer, 2021).

3. Consolidating sustainability management tools and reporting standards

Multinational corporations rely on financial auditing and assurance practices adapted to review performance and impact of sustainability and non-financial reporting. Research based in institutional and legitimacy theories, understanding the resources available to a company, and maturity of sustainability reporting guidelines for independent authorities critically informs how organisations use sustainability management tools (Perego and Kolk, 2012). Continued research on the decision-making factors and influences on organisations' selection and use of sustainability management tools may offer insight to processes for consolidating tools in the future.

As demand for understanding and research on the financial performance impacts of sustainability practices increases, organisations are seeking a consolidated set of standards and guidelines to report their sustainability activities. As evidence of this, the IFRS Foundation is partnering with the CDSB and the VRF to form the International Sustainability Standards Board (ISSB), providing aligned sustainability reporting standards to meet investor expectations (IFRS, 2021). Consolidated reporting standards is a necessary research area to better understand how organisations choose to engage in reporting standards by the types of information they disclose, and whether or not it is validated by some assurance process.

In the next section, we discuss the evolution of sustainability reporting assurance in response to managing the sustainability reporting process enabling transparency in sustainability communications.

4.   Sustainability Reporting Assurance

Mandating organisations to disclose sustainability performance presents a market for assurance tools and services for validating report content and evaluating conformance to sustainable development standards (Grewal and Serafeim, 2020). As more countries mandate public and private organisations to report impact on climate change, there is a rising opportunity for reporting assurance as a service. Similar to management tools like the GRI, TCFD, and BIA, sustainability reporting assurance is seemingly rooted in financial assurance practice, and there is need for a standardised approach (Perego and Kolk, 2012).

Sustainability reporting assurance is a new area of research and practice, with mixed perspectives on the function and impact of how the assurance process adds value to sustainability reporting. Reporting organisations adapt existing financial assurance processes, similar to how reporting standards are based in financial reporting principles. Beyond environmental regulatory compliance, there are currently no standard frameworks or regulation to manage the value of sustainability reporting assurance, which devalues the credibility of existing assurance practices (Grewal and Serafeim, 2020).

There is a difference in how assurance practices are used, which is often based on an organisation's capacity and corporate governance structure. Of the adopting organisations, expanding the remit of traditional financial audit committees or external audit organisations to include sustainability criteria evolves the function reserved for financial auditing practices (Al-Shaer and Zaman, 2018). It is typical of larger, multinational organisations to seek external and third-party assurance services that verify the content and quality of sustainability reports (Perego and Kolk, 2012). While adapting an existing framework from financial auditing principles may help as an initial step, relying on this foundation may create limitations in the comprehensiveness in validating non-financial performance. Research through the theoretical perspectives of stakeholder and legitimacy theories may reveal the benefits and challenges of including non-financial performance, and push to engage other stakeholders like customers or communities in the sustainability management approaches.

External assurance is concerned with an organisation's capacity and corporate governance structure, as well as the priority status of sustainability reporting. The current practices stem from financial assurance and restatement methods and tools, but there are many concessions made in this adaptation

(Michelon, Patten and Romi, 2019). There is a need to explore the value of corporate governance leveraging internal audit resources (Aureli et al., 2020). This research could enhance the framework for implementing sustainable development ideas into existing financial assurance procedures, as well as study how organisations can best employ internal and external resources to create value for stakeholders. Using financial assurance practices is a way to standardise how sustainability management tools are used by organisations, attempting to mitigate variability in report outputs across global, corporate sustainability management tools (Perego and Kolk, 2012). Alignment with sustainable development principles in sustainability report assurance is a risk, as TBL principles are often not well-integrated into traditionally finance-based auditing principles.

Sustainability-reporting assurance services can polarise or homogenise corporate sustainability management tools and practices. A clear, accountable corporate governance structure is key. Segregating the assurance and evaluation functions of sustainability has a negative impact on the sustainability outcomes and the organisation's corporate governance (Al-Shaer and Zaman, 2018). However, including non-financial auditing practices for sustainability reporting improves the credibility of the report content and the brand of the reporting firm (Al-Shaer and Zaman, 2018). External assurance practices can improve report content, yet run the risk of isolating corporate governance from accountability for impact of sustainability performance.

Organisations with a clear corporate governance structure and management approach to sustainability may benefit from external reporting assurance services. The variation of sustainability management tools dilutes the effectiveness of assurance processes and services to validate corporate reporting (Perego and Kolk, 2012). Additional research into the impact of assurance on different types of sustainability management tools may bring clarity to how organisations choose tools to meet stakeholders' needs and improve the content of their reports.

With improvements in the consolidation of sustainability management tools (e.g., as with ISSB) and shareholder and investor demand for transparency and responsibility, sustainability report assurance may be the next solution to a self-created problem. Organisations use many of the sustainability management tools discussed in this chapter, but external assurance providers do not, which creates a misunderstanding of the baseline for conformance and performance evaluation (Miller and Proctor, 2017). Integrating sustainability reporting and consolidating management tools, on the other hand, will increase the alignment of principles used to report and validate any organisation's sustainability performance. In the next section, the need for exploring the links between types of sustainability reporting, quality of assurance services, and implication of meeting stakeholder needs is discussed.

5.   Direction of future research

Further research is needed to understand how sustainability reporting assurance, as a practice and a service, will benefit organisations in measuring the progress of sustainability initiatives. Additionally, there is a need to understand how such practices can be standardised and made accessible as a key resource for all types of organisations publishing sustainability reports as a result of using sustainability management tools. The process to verify sustainability reports is still in its early stages and needs academic- and practitioner-supported development (Boiral, Heras-Saizarbitoria and Brotherton, 2019). There is an opportunity to further explore how standardising organisations' use of sustainability management tools may be of benefit. For example, research into how organisations become aware of sustainability management tools, evaluate and select tools for use, and how organisations communicate use of tools in general business operations to internal and external audiences, could inform how and why some tools are used or others are not used at all.

Another research consideration is the value added of sustainability management tools when used for performance management. This direction of research informs the relationships between an organisation's financial performance, sustainability performance, and longevity. As more standards become available, there is a need to develop assurance methods and frameworks to validate reported content. Continued research into the impact of audited and assured sustainability reporting may refine the understanding of sustainability management practices influencing an organisation's credibility with various stakeholder groups, such as employees, shareholders, customers, community partners, or suppliers throughout the supply chain.

To expand the research outcomes of sustainability reporting and assurance, there is an opportunity to explore concepts using the frameworks of Stakeholder Consultation introduced by Talbot, Raineri, and Daou (2021), and Complexity Theory as presented by Dominici and Roblek (2016). Stakeholder Consultation is an integrative framework bringing to light the influence of stakeholder inputs (consultation) on managerial decisions for using various sustainability management tools and reporting standards (Talbot, Raineri and Daou, 2021). The capability of practitioners using sustainability management tools to successfully engage stakeholders and achieve stakeholder requirements in sustainability management is a significant element in this framework. Further research that examines how practitioners can be most effective in using the tools would be of benefit in this area.

The findings of such a study could lead to the development of reporting standard benchmarks that make reporting more comparable, as well as establish aggregate indicators to effectively monitor progress and impact. Organisations often put more emphasis on disclosing brand-positive information in their corporate reporting, representing aspirational behaviours rather than actionable strategy and measurable contributions (Boiral, Heras-Saizarbitoria and Brotherton, 2019). This can be seen in the above case examples of both M&S and Target by way of which management and assurance tools are used to communicate sustainability performance. Adaptable reporting and assurance methodologies would aid businesses in all sectors and industries by allowing them to present clear proof of their positive and negative societal and environmental impacts by using verified performance indicators and narrative in their reporting.

Comparable sustainability reporting across standards, sectors, and industries could support the development of reporting assurance frameworks. The GRI Standards Assurance service for GRI-framed reports has a loose approach to verifying quality of information over compliance with key terms in reporting, but lacks an approach to critically evaluate or verify performance measures and content of sustainability reports. Although there is an increase in consulting firms that specialise in sustainability reporting and report assurance services in accordance with select reporting standards, their approach is neither regionally nor globally standardised for conformance and compliance to measure the impact or contribution to sustainable development. Further research is therefore needed to understand the links between effectiveness of organisations' sustainability management outcomes and impacts on social and environmental factors external to organisations' operating boundaries.

6.   Conclusion

New standards are continually being developed, posing a capacity challenge for organisations that rely on frameworks like GRI, TCFD, or the SDGs to articulate strategic decisions about sustainability performance. Efforts to consolidate similar reports, such as the IFRS efforts with the ISSB, are an attempt to align resources across sectors and industries to simplify reporting processes.

Reporting assurance is becoming more popular as a means of ensuring accountability and openness, but as a process, it lacks a standardised approach. Consolidated reporting standards, with efforts to normalise integrated reporting, may establish a process replicable in assurance and audit of reports, as well as establish consistent practices to measure the contribution and performance of sustainability management across sectors and industries.

To help organisations convey their commitment to sustainable development, sustainability management tools frequently come in the form of reporting standards or frameworks. The role and impact of the assurance process is a new topic of research, with different perspectives on its function and impact. The sustainability management tools discussed in this chapter provide the foundation to an organisation's building an integrated sustainability strategy. As more organisations publish reports, discussing intents and measures of sustainability initiatives, the demand now turns to verifying contents of reports and tracing performance outcomes to impacts on society.

References

Al-Shaer, H. and Zaman, M. (2018) 'Credibility of sustainability reports: The contribution of audit committees', *Business Strategy and the Environment*, 27(7), pp. 973–986. doi: 10.1002/bse.2046.

Aureli, S. et al. (2020) 'Nonfinancial reporting regulation and challenges in sustainability disclosure and corporate governance practices', *Business Strategy and the Environment*, 29(6), pp. 2392–2403. doi: 10.1002/bse.2509.

Bergman, M. M., Bergman, Z. and Berger, L. (2017) 'An empirical exploration, typology, and definition of corporate sustainability', *Sustainability (Switzerland)*, 9(5), art. 753. doi: 10.3390/su9050753.

Boiral, O., Heras-Saizarbitoria, I. and Brotherton, M. C. (2019) 'Assessing and improving the quality of sustainability reports: The auditors' perspective', *Journal of Business Ethics*, 155(3), pp. 703–721. doi: 10.1007/s10551-017-3516-4.

Camilleri, M. A. (2018) 'Theoretical insights on integrated reporting: The inclusion of non-financial capitals in corporate disclosures', *Corporate Communications*, 23(4), pp. 567–581. doi: 10.1108/CCIJ-01-2018-0016.

CDSB (2019) 'CDSB Framework', (June). https://www.cdsb.net/what-we-do/reporting-frameworks/environmental-information-natural-capital.

Cetindamar, D. and Husoy, K. (2007) 'Corporate social responsibility practices and environmentally responsible behavior: The case of the United Nations Global Compact', *Journal of Business Ethics*, 76(2), pp. 163–176. doi: 10.1007/s10551-006-9265-4.

Chofreh, A. G. and Goni, F. A. (2017) 'Review of frameworks for sustainability implementation', *Sustainable Development*, 25(3), pp. 180–188. doi: 10.1002/sd.1658.

Christofi, A., Christofi, P. and Sisaye, S. (2012) 'Corporate sustainability: Historical development and reporting practices', *Management Research Review*, 35(2), pp. 157–172. doi: 10.1108/01409171211195170.

Costanza, R. et al. (2016) 'Modelling and measuring sustainable wellbeing in connection with the UN Sustainable Development Goals', *Ecological Economics*, 130, pp. 350–355. doi: 10.1016/j.ecolecon.2016.07.009.

Dominici, G. and Roblek, V. (2016) 'Complexity theory for a new managerial paradigm: A research framework', in I. V. Raguz, N. Podrug and L. Jelenc (eds.), *Neostrategic Management*, Springer: Cham, pp. 223–241. doi: 10.1007/978-3-319-18185-1_14.

Eccles, R. G. (2018) 'From climate change the motivation for implementing the TCFD recommendations: The practicality of implementing TCFD', *MIT Sloan Management Review*. http://mitsmr.com/2pbl35s.

Elkington, J. (1994) 'Towards the sustainable corporation: Win–win–win business strategies for sustainable development', *California Management Review*, 36(2), pp. 90–100. doi: 10.2307/41165746.

Elkington, J. (2018) 'Years ago I coined the phrase "triple bottom line": Here's why it's time to rethink it', *Harvard Business Review*. https://hbr.org/2018/06/25-years-ago-i-coined-the-phrase-triple-bottom-line-heres-why-im-giving-up-on-it.

Grewal, J. and Serafeim, G. (2020) 'Research on corporate sustainability: Review and Directions for future research', *Foundations and Trends in Accounting*, 14(2), pp. 73–127. http://www.climateaction100.org/.

GRI (2020) 'Consolidated set of GRI standards'. https://www.globalreporting.org/how-to-use-the-gri-standards/resource-center/.

GRI (2022) *The GRI Perspective: ESG standards, frameworks and everything in between*. https:// www .globalreporting .org/ media/ jxkgrggd/ gri -perspective -esg -standards -frameworks.pdf.

GRI, United Nations Global Compact and World Business Council for Sustainable Development (2016) *SDG Compass: The guide for business action on the SDGs*. http:// sdgcompass .org/ wp -content/ uploads/ 2015/ 12/ 019104 _SDG _Compass _Guide _2015.pdf.

Haller, A., van Staden, C. J. and Landis, C. (2018) 'Value added as part of sustainability reporting: Reporting on distributional fairness or obfuscation?', *Journal of Business Ethics*, 152(3), pp. 763–781. doi: 10.1007/s10551-016-3338-9.

Hamad, S., Draz, M. U. and Lai, F. W. (2020) 'The impact of corporate governance and sustainability reporting on integrated reporting: A conceptual framework', *SAGE Open*, 10(2), pp. 1–15. doi: 10.1177/2158244020927431.

Hayatun, A., Burhan, N. and Rahmanti, W. (2012) 'The impact of sustainability reporting on company performance', *Journal of Economics*. www.industryweek.com.

IFRS (2021) 'IFRS to consolidate CDSB and VFR reporting standards', 3 November. https:// www .ifrs .org/ news -and -events/ news/ 2021/ 11/ ifrs -foundation -announces -issb-consolidation-with-cdsb-vrf-publication-of-prototypes/.

Ike, M. et al. (2019) 'The process of selecting and prioritising corporate sustainability issues: Insights for achieving the Sustainable Development Goals', *Journal of Cleaner Production*, 236, art. 117661. doi: 10.1016/j.jclepro.2019.117661.

Klovienė, L. and Speziale, M. T. (2015) 'Sustainability reporting as a challenge for performance measurement: Literature review', *Economics and Business*, 26. doi: 10.7250/eb.2014.019.

Machado, B. A. A., Dias, L. C. P. and Fonseca, A. (2021) 'Transparency of materiality analysis in GRI-based sustainability reports', *Corporate Social Responsibility and Environmental Management*, 28(2), pp. 570–580. doi: 10.1002/csr.2066.

Marks and Spencer (2021) *Plan A Report 2021*. https:// corporate .marksandspencer .com/sustainability.

Michelon, G., Patten, D. M. and Romi, A. M. (2019) 'Creating legitimacy for sustainability assurance practices: Evidence from sustainability restatements', *European Accounting Review*, 28(2), pp. 395–422. doi: 10.1080/09638180.2018.1469424.

Miller, K. and Proctor, T. Y. (2017) *Current Trends and Future Expectations in External Assurance for Integrated Corporate Sustainability Reporting*. https://www .researchgate.net/publication/321106240.

Nigri, G. and Baldo, M. Del (2018) 'Sustainability reporting and performance measurement systems: How do small- and medium-sized benefit corporations manage integration?', *Sustainability (Switzerland)*, 10(12). doi: 10.3390/su10124499.

Nigri, G., Michelini, L. and Grieco, C. (2017) 'Social impact and online communication in B-Corps', *The Global Journal of Business Research*, 11(3), pp. 87–104.

Orzes, G. et al. (2020) 'The impact of the United Nations Global Compact on firm performance: A longitudinal analysis', *International Journal of Production Economics*, 227, art. 107664. doi: 10.1016/j.ijpe.2020.107664.

Perego, P. and Kolk, A. (2012) 'Multinationals' accountability on sustainability: The evolution of third-party assurance of sustainability reports', *Journal of Business Ethics*, 110(2), pp. 173–190. doi: 10.1007/s10551-012-1420-5.

Sustainability Accounting Standards Board (2017) *Sustainability Accounting Standards Board SASB Conceptual Framework*. https://www.sasb.org/standards/.

Talbot, D., Raineri, N. and Daou, A. (2021) 'Implementation of sustainability management tools: The contribution of awareness, external pressures, and stakeholder consultation', *Corporate Social Responsibility and Environmental Management*, 28(1), pp. 71–81. doi: 10.1002/csr.2033.

Tavares, M. da Costa and Dias, A. P. (2018) 'Theoretical perspectives on sustainability reporting: A literature review', *Accounting from a Cross-Cultural Perspective*. doi: 10.5772/intechopen.76951.

TCFD (2017) *Recommendations of the Task Force on Climate-related Financial Disclosures, Task Force on Climate-related Fiancial Disclosures*. https://assets.bbhub .io/company/sites/60/2020/10/FINAL-TCFD-Annex-Amended-121517.pdf.

The Target Corporation (2021) *2021 Target Corporate Responsibility Report*. https:// corporate .target .com/ sustainability -ESG/ governance -and -reporting/ reporting -progress/archive.

United Nations (2021) *United Nations Global Compact*. https://www.unglobalcompact .org/about.

Villela, M., Bulgacov, S. and Morgan, G. (2021) 'B Corp certification and its impact on organizations over time', *Journal of Business Ethics*, 170(2), pp. 343–357. doi: 10.1007/s10551-019-04372-9.

12   Digital disruption: towards a research agenda for sustainability and business in a digital world

Rory W. Padfield, Alexandra Dales, Jyoti Mishra and Thomas Smith

1. Introduction

> Technology has great potential to help deliver the Sustainable Development Goals, but it can also be at the root of exclusion and inequality. We need to harness the benefits of advanced technologies for all. (UN Secretary-General, Antonio Guterres, at the closing of the 2018 High-level Political Forum on Sustainable Development, United Nations, 2018)

In the past decade, researchers from a range of disciplinary backgrounds have 'turned' towards the digital as an object and subject of inquiry. The emergence of the internet and the rapid innovation in digital technologies has driven research to better understand how the digital re-shapes our world, mediates the production of knowledge, enhances technical efficiencies, and reconfigures relationships with people, power relations and the environment. As Guterres intimates above, it is important to recognise, account for and mitigate against the range of possibilities – both positive and negative – of any kind of techno-logical revolution.

The specific role, scale and means by which businesses and organisations have adopted digital innovations has become a growing focus of interests for the research community. The proliferation of digital goods and services platforms (e.g. eBay, Airbnb, Amazon, Weibo), the rise of national and supra-national data sharing policies (African Union, n.d.; European Commission, 2022), and the emerging policy discourse on digital technology's role in supporting public needs and services (e.g. health care, education, agriculture) all underscore the timeliness of business and organisational themed research in this field. As Feroz, Zo and Chiravuri (2021) observe, the digital transformation of society has major implications for businesses, markets, and industries, which in turn, leads to new opportunities for business model creation or the streamlining of existing ones. Research, therefore, plays a key role in informing the direction of these new developments.

One key area of inquiry within this field relates to the sustainability of digital innovations in a business and organisational context; currently, there is considerable uncertainty about the sustainability impact of digital innovations across different types and scales of business, as well as in different geographies. Accordingly, in this chapter we seek to first define and describe digital innovations in a business context, followed by a critical review of digital innovations from a sustainability perspective. We conclude the chapter by identifying a research agenda in the field, which includes suggestions on key research methods.

2.   Definitions and applications

A starting point in any synthesis of the digital innovation literature is a short reflection on the key definitions and terminology applied in this context. In addition to the conventional terminology of 'hardware' and 'software' – terms that refer to the physical computing infrastructure, and the software applications and underpinning code and algorithms – there are a number of frequently used terms that require explanation before critical analysis of their contribution to the field of sustainability and business.

One of the most commonly used terms in this field is the Internet of Things (IoT). Haller et al. (2009) define IoT as:

> ...a world where physical objects are seamlessly integrated into the information network, and where the physical objects can become active participants in business processes. Services are available to interact with these 'smart objects' over the Internet, query their state and any information associated with them, taking into account security and privacy issues. (Haller et al., 2009: 15)

In effect, the IoT enables computer devices (smartphones, smart home devices, and connected industrial equipment), to 'talk to one another' and transfer information between them without the need for a physical human presence. The number of connected devices worldwide is expected to amount to 30.9 billion units by 2025, which represents a considerable increase from the 13.8 billion units in 2021 (Vailshery, 2022).

Another frequently used term is Web 2.0, which refers 'to the array of contemporary web-technologies that allow users to interactively engage people and media content in different ways, such as participatory information sharing' (Ballew et al., 2015: 10621). Examples include social media and information sharing platforms, for example Meta (Facebook, Instagram, WhatsApp), WeChat and Wikipedia. These platforms allow users to consume, create, refine and share data to a wide audience.

The information or data moving between users in Web 2.0 is often referred to as Big Data. Sivarajah et al. (2017) argue that Big Data has seven main characteristics (also known as the '7 Vs'): veracity, volume, velocity, variety, visualisation, variability, and value. In the context of businesses, Big Data ecosystems describe 'the data environment supported by a community of interacting organisations and individuals' (Curry, 2016: 33). The '7 Vs' of Big Data is thus an important concept when considering data movements between individuals, businesses and organisations and how they evolve at any one moment in time.

The digital economy refers to 'economic output derived solely or primarily from digital technologies with a business model based on digital goods or services' (Bukht and Heeks, 2017: 13). Digital and platform services, such as Amazon, Uber and Google represent lead firms in the global digital economy. Mack and Veil (2017) conceptualise digital platform services as multi-sided platforms (MSPs) characterised by a complex marketised structure, which can influence the business models firms choose to develop. Industry 4.0 is understood as a new paradigm of smart and autonomous manufacturing, which aims to integrate manufacturing operations systems with communication, information and intelligence technologies (Wang et al., 2017; Jeschke et al., 2017).

Finally, Artificial Intelligence (AI) refers to machines performing various cognitive functions associated with humans, including perceiving, reasoning, learning, and interacting (Rai et al., 2019). The fundamental principle of AI is machine learning, that is, the ability of a computer to improve its own abilities continuously (Garbuio and Lin, 2019) and typically, they support business operations with assisted, augmented and autonomous intelligence.

Table 12.1 summarises the range of digital innovations that are re-shaping the business landscape with example technologies provided.

Table 12.1   Examples of digital technology innovation and their application within a business context

Types of Technology	Description	Business Application	Examples
Social media	Online platform or website that allows users to create and share information (e.g. text, images, videos) to wide audience.	Mainly used for projecting corporate social responsibility activities; data sharing on company initiatives related to sustainability, posts related to making products more recyclable, impact on consumption habits.	Nestlé's post on no plastic should end up in landfill; IKEA's fortune favours the frugal advertisement on YouTube; Interface
Mobile technology devices and software	A device which is typically characterised by mobility, small form factor and communication functionality, and focuses on handling a particular type of information and related tasks, e.g. smartphone, tablet (Tselios et al., 2008).	Provides ubiquity, accessibility and immediacy. Increases flexibility at work, such as hybrid working.	Teams, Slack, Zoom, Trello, Workful, Gusto; Tap; Joule Bug
Analytics and visualisation	Capturing, storing, analysing, sharing and linking vast amounts of data.	Transparency of business operations, including monitoring of risk in supply chains; communication of business activity to external audience; multi-layered analysis of sensor data in a city setting; people analytics, employees wellbeing.	Sime Darby's palm oil traceability tool (Cross Check); Cool Farm Tool; Winnow's food waste minimising technology

Types of Technology	Description	Business Application	Examples
Cloud computing	Provisioned IT services accessed from a cloud-computing provider and real-time communication software. Enables high quality business-to-business communication and a convenient, on-demand network access to a shared pool of configurable computing resources (e.g. networks, servers, storage, applications, and services) rapidly provisioned and released with minimal management effort or service provider interaction (Bello et al., 2021).	Applied in wide range of businesses, sectors and organisational types, including for data analytics. Impacts of COVID-19 have underscored the value of virtual computing infrastructure to many organisations (e.g. service industries, higher education) to facilitate working from home practices.	Use of Google Apps by SMEs; Amazon Web Services for agricultural applications; Zoom for business–business communication
Augmented, virtual and mixed reality	Integration of the actual world with digital information about it.	Engaging with customers and employees, simulating working in dangerous environments.	AR-labelling for product marketing; Smart glasses to identify products in supermarket context; Steamed Egg; VR for safety training

Types of Technology	Description	Business Application	Examples
Smart sensors	Devices that are sensitive to, detect, measure and convert stimuli from the physical environment into a usable electrical signal. Typically connected to a micro-controller, smart sensors are capable of running algorithms and storing data, and send data or receive commands using digital protocols and data formats (Sagar et al., 2018).	Wide variety of applications, including health care, manufacturing, city design and monitoring, building management, security.	Siemens' Smart City Concept; Manufacturing of photovoltaics; Newcastle Urban Observatory
Satellite & airborne technology	Broad category including satellite and airborne technology, such as Unmanned Aerial Vehicles/Systems (UAV/UASs) (also known as 'drones'). Devices can generate high-resolution granular digital videos, images and maps, and undertake a growing number of in-situ tasks. Digital outputs can be made widely accessible online via usable files types, e.g. jpeg, shapefiles.	Industrial inspections, security surveillance, precision agriculture, filming, broadcasting, delivery and logistics (Giones and Brem, 2017).	Crop and biodiversity surveillance in Southeast Asia; Tree planting in Canada; Goods delivery in remote parts of India

Types of Technology	Description	Business Application	Examples
Sharing Economy Platforms	Business model boosted by internet to enable information and skills to be shared or exchanged (Botsman and Rogers, 2010; Mack and Veil, 2017). It is characterised by temporary access, redistribution of goods or less tangible assets such as space, time.	Wide variety of applications, such as in the accommodation sector whereby users can rent a property via interaction with owner on a purposed built digital platform. People become less dependent on ownership.	BlaBlaCar (car pool); Airbnb (accommodation); OBO (bike sharing); TooGoodToGo (food waste redistribution); Couchsurfing (hospitality exchange)
Blockchain	A distributed, public or private digital ledger that uses cryptography and authentication technology to record business transactions (e.g. sales, quality) across IT networks to prevent records from being altered retrospectively (Lakkakula et al., 2020).	Supply chain transparency from farm to fork; verifiable authenticity.	IBM Food Trust (https://www.ibm.com/uk-en/blockchain/solutions/food-trust)

3.   Critical perspectives of digital innovation in sustainability and business

Focusing on two specific sustainability challenges associated with digital tools and technologies, the next section critically analyses the environmental and social impact of digital innovations with particular reference to published research in the sustainability and business fields.

3.1   Resource efficiency gains but at what cost?

Proponents of various types of digital innovation point heavily towards a broad range of environmental benefits (Bai et al., 2020; Gensch et al., 2016; Jeschke et al., 2017; Liu et al., 2020; Wang et al., 2017). These benefits primarily focus on the sharing of resources – for example private vehicles and accommodation, bicycles, electrical goods – and the efficiency gains from improved management and engineering systems. The manufacturing sector is a case in point; traditional manufacturing systems are characterised as highly damaging to the environment (e.g. contributing to climate change from historical and present-day greenhouse gas emissions, high levels of resource use) and facing various social issues in the sites of production, such as poverty and inequality for factory workers (Bai et al., 2020). Digital manufacturing on the other hand, known as Industry 4.0 or the fourth industrial revolution, is 'enabling ever-higher levels of production efficiencies' (Bai et al., 2020: 107776), which in turn leads to 'high-efficiency, high-quality, and cost-effective products and services' (Liu et al., 2020: 107889). Examples include 'intelligent manufacturing industries' in Germany and the US, which use complex and vast digital data sets from various automated systems to support the design and manufacture of aircraft engines (Liu et al., 2020).

Efficiency gains brought about by digital innovation are not isolated to the manufacturing sector but experienced across a range of business sectors, including financial services, engineering and construction, and medical. Gensch et al. (2016) refer to the implementation of a broad-based 'digitalisation strategy' across a wide variety of organisational types. Underpinned by data centres (for storage and processing) and telecommunication networks (for transmission and communication purposes), this type of greening strategy has supported a working-from-home culture across many parts of the world – which enables reduction in environmental and social impacts related to commuter travel – and more comprehensive sharing of digital resources between and within organisations. For example, across different parts of the world a large global team of architects and engineers can make decisions and modify plans for the design and construction of a building. Not-for-profit organisations, such as charities, social enterprises and non-governmental organisations (NGOs) are also adopting digital technologies to expand or strengthen their work and operations. Box 12.1 describes the way an NGO has successfully employed drone imagery to tackle deforestation in Malaysia.

Box 12.1 Application of drone technology to halt deforestation in Malaysia

The widespread availability of drones combined with the accessibility of their user-friendly digital technologies, such as automated fly-by-GPS piloting and gyro-stabilised photography, has facilitated a significant socio-technical transition in the environment sector. Unmanned Aerial Systems (UAS) – a collective term for various aerial vehicles and associated technologies – have become notable 'information generators' for environmental scientists (Avron, 2017). Their use by NGOs, charities and campaign groups, for environmental advocacy and policing is revolutionising approaches to conservation.

A common tactic used by industry to lobby for agricultural conversion of forests in tropical countries is to claim that the forest in question is 'heavily degraded' and thus provides few benefits in terms of biodiversity, carbon storage and other ecosystem services (Thompson et al., 2013). This was the case for a forest in northern Selangor, Peninsular Malaysia. Roadside photographs presented in favour of agricultural development showed a shrubby landscape with trees in poor health. A local NGO, the Global Environment Centre (GEC) was able to rapidly deploy their UAS and capture aerial photographs showing that beyond the road, there was a highly biodiverse and intact secondary forest rich with endemic tree species. Furthermore, GEC argued that if developed, the plantations would be prone to flooding due to land subsidence. The NGO presented compelling evidence using drone imagery of a neighbouring oil palm plantation under water with the intact healthy forest on the other side of the road (see Figure 12.1).

Coupled with evidence from longer-term camera trap monitoring in nearby forests, GEC were able to persuade the state government officials to halt the development. Prior to the availability of drone technology, responding to announcements of development in forests was a slow process, often requiring expedition surveys run by experienced teams. UASs have empowered environmental NGOs with a rapid response and a means to gather persuasive evidence in favour of conservation.

Figure 12.1 Aerial photograph captured in Peninsular Malaysia by a drone showing (on the right) the edge of a peat swamp forest proposed for development with an intact, diverse and dense canopy; and (on the left) a neighbouring palm oil plantation under flood conditions

Notwithstanding the reported environmental gains, researchers have started to unravel some of the negative impacts associated with digital innovations. The rise and availability of digital technologies across the world has led to an associated sustainability phenomenon – electronic waste (or E-waste). E-waste is defined as 'all types of electrical and electronic equipment and its parts that have been discarded by the owner as waste without intention of re-use' (Kumar et al., 2017: 33). The notion of 'product obsolescence' – understood as a perceived or real loss of product value typically triggered by diminished functionality (den Hollander et al., 2017) – and the ever-increasing consumer demand for the latest electronic models and devices driven by the power of digital marketing has led to a toxic accumulation of E-waste. Reflecting on the fast speed of information communication against the 'slow violence' of E-waste impacts, LeBel (2016) observes that:

> …Policies of planned obsolescence dictate that computers and other ICTs be 'death dated' so that existing models are not backwards compatible. New, 'improved', and faster models are marketed every few months and, as a result, functioning ICTs are routinely upgraded and discarded. Planned obsolescence accelerates the accumulation of trashed electronics. (LeBel, 2016: 301)

It is estimated that the UK produced 505,445 tonnes of E-waste in 2019, which equates to 23.9 kg per person (House of Commons, 2020). Notwithstanding the sheer scale of this amount of waste, the global geography of E-waste circuitry is of equal concern. Reinforcing postcolonial inequalities between the Global South and Global North, countries such as the Philippines, Ghana and Nigeria have become 'digital dumping grounds' (Akese and Little, 2018) for E-waste from the Global North. Empirical studies of E-waste governance in Ghana observe unimaginable working conditions for workers who aim to salvage usable parts for resell, many of whom are exposed to harmful pollutants from the burning of E-waste (Akese and Little, 2018; Sovacool, 2019).

A less visible social impact of global E-waste circuitry is digital content moderation. In E-waste processing sites, digital content stored on the hardware in these sites (e.g. hard drives, laptops, PCs, smartphones) is often moderated before being resold or repurposed. Wan (2021) observes that content moderators are often young workers searching for jobs in the South and Southeast Asian technology sector. In the Philippines, Roberts (2016) describes the moderation of disturbing digital content – material designed to shock or is offensive by nature, such as pornography, the depiction of adult and child abuse, and uncensored material from war zones – leaving workers vulnerable to mental illnesses and secondary psychological trauma. In her investigations, Roberts identifies Canada as the source of the disturbing digital content moderated by Filipino workers. Wan (2021: 2646) argues that 'without proper safeguards and regulations to tackle the issues of waste processing…the rhythms and durations of colonial violence will continue to haunt our globalized world'.

The impacts of E-waste are not only felt at the point of re-use or disposal. A UK government report highlighted the embedded carbon emissions in electronic devices, as well as the high levels of resource extraction to build electronic devices, including critical resource materials (CRMs) (House of Commons, 2020). For example, indium is a CRM used in touchscreens and solar panels, and tantalum is used in a range of applications from mobile phones to wind turbines (House of Commons, 2020). These materials are commonly found in countries in the Global South where weak regulatory frameworks can fail to protect the rights and the welfare of workers (George, 2019).

The mining of precious metals can also intensify conflict between groups, which further exposes local people and communities to threats and potential violence. For instance, the gold, tin, tungsten and tantalum necessary for the production of electronics goods are known as 'conflict minerals'. The collapse of an illegal gold mine in Indonesia in 2019 – gold is a rare metal often used in smartphones – which led to the deaths of over 100 miners illustrates the vulnerability of workers and local communities in the global electronics sector. The mining and sale of these minerals are linked with human rights abuses (e.g. funding killings, violence, rape) in the Democratic Republic of Congo (DRC) and other conflict zones (House of Commons, 2020). Similarly, cobalt used in lithium-ion batteries, is associated with child labour in the DRC (House of Commons, 2020). Health researchers have highlighted the exposure of workers to pollution in heavy metal mining for the electronics manufacturing industry (Cai et al., 2018), as well as the rare earth mining industry (Shin et al., 2019).

Next, the chapter moves onto a critical analysis of the sharing economy with the objective to examine a number of social, economic and environmental impacts.

3.2   The sharing economy – social and economic empowerment but for whose benefit?

The sharing economy is regarded as a new model of consumption, which relies on peer-to-peer interaction via digital platforms. As shown in Table 12.1, sharing economy platforms provide a wide range of economic and social interactions between digitally connected individuals, and in many cases without the need for a third party or intermediary. As compared with more conventional models of consumption, the emergence of the sharing economy has led to 'sharing' rather than 'possession' of goods and services, and placing consumers/users in direct contact with sellers or other actors. Examples of digital platforms include the creative commons (e.g. Wikipedia), intermediary and direct trading activities (e.g. Uber, Just Eat, Amazon, Airbnb), and crowd-funding activities (e.g. Just Giving, GoFundMe).

The sharing economy has become big business. In the UK, it is estimated that the country's sharing economy will be worth £140 billion by 2025 (CBI, 2021), and the direct economic benefits are regarded as indisputably positive (Frenken and Schor, 2017). The resulting transactions and exchange of goods and services has led to an increase in personal wealth and the creation of new business opportunities. In turn, this has created a degree of economic and social empowerment for individuals, especially in circumstances where lenders had limited access to markets in the first place.

The dynamism of the sharing economy concept also allows individuals and sharing organisations to produce new goods and services and access new markets at short notice. Sharing Economy UK, the trade body for sharing organisations in the UK reports the varied ways their members have adapted to the economic impacts of the coronavirus pandemic. This includes making available cleaning and disinfection guides for passengers using car sharing apps, and the creation of new mobile cleaning companies to provide hosts with access to on-demand, and affordable vehicle sanitisation services (Sharing Economy UK, no date).

The notion of sharing rather than possessing goods and services has a number of sustainability benefits. Early research has previously argued that sharing instead of owning goods and services may reduce consumption-induced resource depletion (Schor and Wengronowitz, 2017). In short, fewer resources are required for the generation of new materials since existing goods and services are shared between users. Despite the considerable potential and realised benefits of the sharing of resources, researchers have also pointed out the potential pitfalls in this argument (Frenken and Schor, 2017); namely, the 'rebound effect'. The rebound effect implies that benefits generated from improved technology (e.g. access to goods and services via digital platforms) are off-set by a change in consumer behaviour, which in turn leads to further consumption of those goods and services (ESRC, 2021). Whilst the collection of meaningful data on this topic is fraught with problems and complexity, studies investigating the extent of the rebound effect across different sharing platforms would be beneficial to the current sharing economy discourse.

From a social perspective, sharing platforms have been shown to create and strengthen 'social ties' between individuals, groups of people and within organisations. For example, crowdfunding platforms can help to raise economic capital via social networking, which in turn can support individual or business needs. In another example, hosts on Airbnb have been shown to widen their own social networks after socialising and making friends with their guests (Schor, 2015).

Despite a range of sustainability benefits, researchers have identified a number of negative impacts associated with the sharing economy. The lack of regulatory frameworks has left workers exposed to exploitation and discrimination. Famously, Morozov (2013) has argued that the sharing economy 'is a form of "neo-liberalism on steroids" which commercialises aspects of life previously beyond the reach of the market' (Martin, 2016: 149). In this way, research has shown how hosts and lenders of specific goods and resources in the sharing economy seem to be in a race to the bottom in terms of working conditions (Henten and Windekilde, 2016; Murillo et al., 2017). High levels of social disruption can also occur, as reported in Amsterdam where a growth in Airbnb properties led to exorbitant rental prices, which in turn forced local families to relocate out of the city (van der Zee, 2016). Thus, opportunities exist to examine further the empirical sustainability impacts of sharing business models and consider what types of business models and regulations would balance out these competing demands. A good starting point for an analysis of business model innovation in the sharing economy space is the work by Ciulli and Kolk (2019), who developed a typology of business model innovation for the sharing economy.

4.   A research agenda for business and sustainability in a digital world

Building on our review of key sustainability issues in terms of business, sustainability and digital innovation, the following discussion sets out a specific research agenda on this topic. Suggestions for research methodology have also been included.

4.1   Sustainability impact studies

The rapid pace of digital technology innovation and implementation makes it challenging to evaluate the sustainability impacts before any intended or unintended outcomes are fully realised. Critical evaluations of the sustainability impacts of new technologies are thus a good starting point, particularly in relation to effects on priority concerns, such as resource depletion, climate change and impacts on vulnerable and marginalised groups, for example minorities, low-income communities and smallholder farmers. Such studies will generate primary empirical knowledge about the innovations themselves, which in turns feed into problematisation and theorisation of these innovations. The outcomes of these studies can contribute towards public and private policy discourse on the merits on these new and emerging digital innovations.

The highly specialised nature of some of these technologies lends itself to studies that are interdisciplinary (e.g. working across disciplinary boundaries) and transdisciplinary (e.g. working with those who apply and are affected by the technology). For instance, researchers from business studies or sustainability backgrounds may need to develop their own digital literacy skills (e.g. coding skills such as Python, R and data analytics) or consider partnerships with researchers with these specific knowledge sets. Digital literacy is likely to be an important part of future studies as researchers aim to understand the underlying make-up, data assumptions and functionality of an innovation, as well as how to measure aspects of the technology. Established social science methods common to sustainability and business research (e.g. interviews, focus groups, life cycle assessment, social impacts assessment, livelihood, and wellbeing assessment methods) have a role in terms of the environmental and social impact analysis. Research teams with a balance of sustainability, business and digital technological background are thus ideal, as well as those who manage to engage industry partners applying such innovations, e.g. software firms, private industry, and government agencies.

4.2 Organisation and regulatory research

Since digital innovation creates the space for new business models (Feroz et al., 2021), the next part of the research agenda focuses on the primary users of these innovations, namely, businesses and organisations. Consequently, this agenda draws on a strong tradition of sustainable business model analysis, organisational change for sustainability, and corporate sustainability fields (Adams et al., 2016; Bocken et al., 2014). This is to determine the extent to which digital innovations challenge existing business paradigms whether for better or worse, and for whom. Future studies could aim to explore the drivers and barriers to innovation adoption, as well as broader institutional constraints and opportunities as experienced by the businesses and organisations themselves. Regulatory focused studies could also play a role, particularly in view of observations regarding the lack of satisfactory worker safeguards in various sharing economy contexts (Murillo et al., 2017). Importantly, these investigations should aim to understand how business models interact with concepts and practices of sustainability within businesses and organisations at different scales. In addition to conventional social science methods for data collection, researchers should aim to foster strong networks with businesses and organisations to support their research aims in this particular theme.

4.3 Digital innovation in the Global South

To date, studies in the digital innovation, business and sustainability field have paid scant attention to the applications and impacts of digital innovations in the Global South. While a small number of studies have started to examine the issue of E-waste in parts of West Africa and Southeast Asia, as discussed above, there are a wide variety of continuing knowledge gaps that require investigation. Accordingly, in this part of the research agenda we recommend a comprehensive analysis of digital innovation within the context of the Global South. This could involve meta-analyses and comparative country analyses, and in-depth empirical research in individual communities. The agenda should consider the Global South not simply as *receivers* of technology but also as *innovators*. Nascent research has underlined the scale and levels of innovation in regions of the Global South, such as digital health care provision in Africa (Olu et al., 2019) and agricultural digital innovation in India (Boettiger and Sanghvi, 2019), as well as research observing the digital marginalisation of communities in the broader Global South (Haenssgen, 2018). Within this agenda there is also scope to apply a critical social theoretical lens (e.g. postcolonialism, political ecology) to investigate patterns of inequality and consider fairer, more just digital innovations.

Methods supporting these studies include conventional social sciences methodologies (as described in section 4.1) but could also comprise more participatory approaches, such as ethnography to determine the lived experience of communities vis-à-vis digital technology and innovation. Meta-analysis studies would benefit from expertise in quantitative approaches, such as statistical analyses, computational mapping and Geographic Information Systems (GIS).

4.4   'Digital storytelling' for sustainable business

The final part of our proposed research agenda highlights opportunities for studies within the theme of digital storytelling. Over the past decade, various businesses and organisations have explored creative ways to communicate aspects of their operations and supply chains for public consumption. Moving beyond conventional corporate reporting activities, an increasing number of companies are developing online interactive tools and resources, which allow the public and other key stakeholders to imagine the scale, scope and impacts of business operations. An example of this type of digital storytelling is Marks and Spencer's (2022) 'Interactive Supply Chain Map' tool, which displays various data points on commodity type, impacts and geographies. As digital storytelling continues to play a prominent role in corporate sustainability communication, opportunities exist for researchers to examine critical issues of supply chain transparency and traceability, the sustainability outcomes of such tools, data ownership and privacy, and to scrutinise the extent to which these tools genuinely influence the behaviour of external stakeholders.

Within this research agenda there are also opportunities to study the role digital specialists in the creative sector (e.g. augmented and virtual reality, media, and film) can play in sustainability communication for businesses and organisations, and society more broadly. In the last decade, creative sector firms have started to apply their unique skills and capabilities to the theme of sustainability. Recent examples include the 'Green Planet AR Experience' (BBC Studios, 2022) and 'Flood' (XR Stories, 2022), both public events used augmented reality technologies to give audiences an immersive and digitally dynamic perspective of themes such as biodiversity, sustainability impacts, and climate change. In view of the considerable potential of the creative sector in sustainability storytelling, researchers with cross-disciplinary epistemological and methodological backgrounds (e.g. communication and media studies, augmented and VR technology studies, sustainability and business studies) are well placed to study this emerging sub-theme.

To conclude, in this chapter we have defined the key terminology associated with digital technology and innovation, and provided some examples of their application in various business contexts. We have summarised some of the key sustainability issues – both positive and negative – associated with digital technology and innovations, with a particular focus on the challenges of resource efficiency and the sharing economy. Finally, to help develop and move this field of study forward we have identified four themes of research with reflections on appropriate methodologies underpinning these themes.

References

Adams, R., Jeanrenaud, S., Bessant, J., Denyer, D. and Overy, P. (2016) Sustainability-oriented innovation: A systematic review, *International Journal of Management Reviews*, 18, pp. 180–205

African Union (n.d) The digital transformation strategy For Africa (2020–2030) [online] https:// au .int/ sites/ default/ files/ documents/ 38507 -doc -dts -english .pdf (Accessed: 01/12/2021)

Akese, G. and Little, P. (2018) Electronic waste and the environmental justice challenge in Agbogbloshie, *Environmental Justice*, 11 (2), pp. 1–14

Avron, L. (2017) 'Governmentalities' of conservation science at the advent of drones: Situating an emerging technology, *Information & Culture*, 52, pp. 362–383

Bai, C., Dallasega, P., Orzes, G. and Sarkis, J. (2020) Industry 4.0 technologies assessment: A sustainability perspective, *International Journal of Production Economics*, 229, 107776

Ballew, M., Omoto, A. and Winter, P. (2015) Using Web 2.0 and social media technologies to foster proenvironmental action, *Sustainability*, 7, pp.10620–10648

BBC Studios (2022) The Green Planet AR Experience powered by EE 5G to launch in London February 2022 [online] https:// www .bbcstudios .com/ news/ the -green -planet -ar -experience -powered -by -ee -5g -to -launch -in -london -february -2022/ (Accessed: 20/03/2022)

Bello, S., Oyedele, L., Akinade, O., Bilal, M., Delgado, J., Akanbi, L., Ajayi, A. and Owolabi, H. (2021) Cloud computing in construction industry: Use cases, benefits and challenges, *Automation in Construction*, 122, 103441

Bocken, N., Short, S., Rana, P. and Evans, S. (2014) A literature and practice review to develop sustainable business model archetypes, *Journal of Cleaner Production*, 65, pp. 42–56

Boettiger, S. and Sanghvi, S. (2019) How digital innovation is transforming agriculture: Lessons from India. McKinsey and Company [online] https://www.mckinsey .com/ industries/ agriculture/ our -insights/ how -digital -innovation -is -transforming -agriculture-lessons-from-india (Accessed: 01/05/2022)

Botsman, R. and Rogers, R. (2010) *What's Mine is Yours*, London: Collins

Bukht, R. and Heeks, R. (2017) Defining, conceptualising and measuring the digital economy. Centre for Development Informatics, Global Development Institute, University of Manchester

Cai, Y., Li, F., Zhang, J. and Wu, Z. (2018) Occupational health risk assessment in the electronics industry in China based on the occupational classification method and EPA model, *International Journal of Environmental Research and Public Health*, 15, 2061

Ciulli, F. and Kolk, A. (2019) Incumbents and business model innovation for the sharing economy: Implications for sustainability, *Journal of Cleaner Production*, 214, pp. 995-1010

Confederation of British Industry (CBI) (2021) Sharing Economy UK trade body to join CBI [online] https://www.cbi.org.uk/media-centre/articles/sharing-economy-uk-trade-body-to-join-cbi/ (Accessed: 15/12/2021)

Curry, E. (2016) The big data value chain: Definitions, concepts, and theoretical approaches. In J. Cavanillas, E. Curry and W. Wahlster (eds), *New Horizons for a Data-Driven Economy A Roadmap for Usage and Exploitation of Big Data in Europe*, Springer: Cham, Switzerland, pp. 29-37

den Hollander, M., Bakker, C. and Hultink, E. (2017) Product design in a circular economy: Development of a typology of key concepts and terms, *Journal of Industrial Ecology*, 21 (3), pp. 517-525

ESRC (2021) Rebound effect [online] https:// webarchive .nationalarchives .gov .uk/ ukgwa/ 20210901141238/ https:// esrc .ukri .org/ about -us/ 50 -years -of -esrc/ 50 -achievements/the-rebound-effect/ (Accessed: 11/03/2021)

European Commission (2022) A European strategy for data [online] https:// digital -strategy.ec.europa.eu/en/policies/strategy-data (Accessed: 21/03/2022)

Feroz, A.K., Zo, H. and Chiravuri, A. (2021) Digital transformation and environmental sustainability: A review and research agenda, *Sustainability*, 13, 1530. https:// doi .org/10.3390/su13031530

Frenken, K. and Schor, J. (2017) Putting the sharing economy into perspective, *Environmental Innovation and Societal Transitions*, 23, pp. 3-10

Garbuio, M. and Lin, N. (2019) Artificial intelligence as a growth engine for health care startups: Emerging business models, *California Management Review*, 61 (2), pp. 59-83

Gensch, C.O., Prakash, S. and Hilbert, I. (2016) Is digitalisation a driver for sustainability? In T. Osburg and C. Lohrmann (eds), *Sustainability in a Digital World*, Springer: Cham, Switzerland, pp. 117–129

George, K. (2019) The tech industry has a serious sustainability problem; and it's going unchecked [online] https://www.huckmag.com/art-and-culture/tech/the-tech -industry-has-a-serious-sustainability-problem/ (Accessed: 03/01/2022)

Giones, F. and Brem, A. (2017) From toys to tools: The co-evolution of technological and entrepreneurial developments in the drone industry, *Business Horizizon*, 60 (6), pp. 875-884

Haenssgen, M. (2018) The struggle for digital inclusion: Phones, healthcare, and marginalisation in rural India, *World Development*, 104, pp. 358-374

Haller, S., Karnouskos, S. and Schroth, C. (2009) The internet of things in an enterprise context. pp. 14-28 In *Future Internet Symposium*, First Future Internet Symposium Vienna, Austria, 28-30 September

Henten, A. and Windekilde, I. (2016) Transaction costs and the sharing economy, *Info*, 18 (1), pp. 1-15.

House of Commons (2020) Environmental Audit Committee: Electronic waste and the circular economy [online] https://committees.parliament.uk/publications/3675/documents/35777/default/ (Accessed: 27/12/2021)

Jeschke, S., Brecher, C., Song, H. and Rawat, D. (2017) *Industrial Internet of Things*, Springer: Cham, Switzerland

Kumar, A., Holuszko, M. and Espinosa, D. (2017) E-waste: An overview on generation, collection, legislation and recycling practices, *Resources, Conservation and Recycling*, 122, pp. 32-42

Lakkakula, P., Bullock, D. and Wilson, W. (2020) Blockchain technology in international commodity trading, *The Journal of Private Enterprise*, 35(2), pp. 23-46

LeBel, S. (2016) Fast machines, slow violence: ICTs, planned obsolescence, and e-waste, *Globalizations*, 13 (3), pp. 300-309

Liu, Y., Zhu, Q. and Seuring, S. (2020) New technologies in operations and supply chains: Implications for sustainability, *International Journal of Production Economics*, 229, 107889

Mack, O. and Veil, P. (2017) Platform business models and internet of things as complementary concepts for digital disruption. In Anshuman Khare, Brian Stewart and Rod Schatz (eds), *Phantom Ex Machina Digital Disruption's Role in Business Model Transformation*, Springer: Cham, Switzerland, pp. 71-86

Marks and Spencer (2022) Interactive supply chain map [online] https://interactivemap.marksandspencer .com/ ?sectionPID = 5c61 439cc6fe1b a1f45131e0 (Accessed: 21/03/2022)

Martin, C. (2016) The sharing economy: A pathway to sustainability or a nightmarish form of neoliberal capitalism?, *Ecological Economics*, 121, pp. 149-159

Morozov, E. (2013) The 'sharing economy' undermines workers rights [online] http://evgenymorozov.tumblr.com/post/64038831400/the-sharing-economy-undermines-workers-rights (Accessed: 11/10/2021)

Murillo, D., Buckland, H. and Val, E. (2017) When the sharing economy becomes neoliberalism on steroids: Unravelling the controversies, *Technological Forecasting and Social Change*, 125, pp. 66-76

Olu, O., Muneene, D., Bataringaya, J.E., Nahimana, M-R., Ba, H., Turgeon, Y., Karamagi, H.C. and Dovlo, D. (2019) How can digital health technologies contribute to sustainable attainment of universal health coverage in Africa? A perspective, *Frontiers in Public Health*, 7, 341. doi: 10.3389/fpubh.2019.00341

Rai, A., Constantinides, P. and Sarker, S. (2019) Editor's comments: Next-generation digital platforms: Toward human–AI hybrids, *MIS Quarterly*, 43(1), pp. iii-ix

Roberts, S. (2016) Digital refuse: Canadian garbage, commercial content moderation and the global circulation of social media's waste, *Media Studies Publications*, 14

Sagar, S., Lefrançois, M., Rebaï, I., Khemaja, M., Garlatti, S., Feki, J. and Médini, L. (2018) Modeling Smart Sensors on top of SOSA/SSN and WoT TD with the Semantic Smart Sensor Network (S3N) modular Ontology. ISWC (2018) 17th Internal Semantic Web Conference, October 2018, Monterey, United States, pp. 163-177

Schor, J. (2015) The sharing economy: Reports from stage one. Report, Boston College

Schor, J. and Wengronowitz, R. (2017) The new sharing economy: Enacting the eco-habitus. In H. Brown, M. Cohen and P. Vergragt (eds), *Sustainable Consumption and Social Change*, Routledge: London, pp. 25-42

Sharing Economy UK (no date) The UK's enduring sharing economy [online] https://www.sharingeconomyuk.com/ (Accessed: 12/11/2021)

Shin, S.-H., Kim, H.-Y. and Rim, K.-Y. (2019) Worker safety in the rare earth elements recycling process from the review of toxicity and issues, *Safety and Health at Work*, 10, pp. 409–419

Sivarajah, U., Kamal, M.M., Irani, Z. and Weerakkody, V. (2017) Critical analysis of Big Data challenges and analytical methods, *Journal of Business Research*, 70, pp. 263–286

Sovacool, B.K. (2019) Toxic transitions in the lifecycle externalities of a digital society: the complex afterlives of electronic waste in Ghana, *Resources Policy*, 64, 101459

Thompson, I. D., Guariguata, M. R., Okabe, K., Bahamondez, C., Nasi, R., Heymell, V. and Sabogal, C. (2013) An operational framework for defining and monitoring forest degradation, *Ecology and Society*, 18(2), 20

Tselios, N., Papadimitriou, I., Raptis, D., Yiannoutsou, N., Komis, V. and Avouris, N. (2008) Designing for mobile learning in museums. In J. Lumsden (ed.), *Handbook of Research on User Interface Design and Evaluation for Mobile Technology*, IGI Global: Hershey, PA, pp. 252–268

United Nations (2018) Secretary-General's remarks at closing of High-Level Political Forum on Sustainable Development [online] https://www.un.org/sg/en/content/sg/statement/2018-07-18/secretary-generals-remarks-closing-high-level-political-forum (Accessed: 22/04/2022)

Vailshery, L. (2022) Number of Internet of Things (IoT) connected devices worldwide from 2019 to 2021, with forecasts from 2022 to 2030. Statista [online] https://www.statista.com/statistics/1183457/iot-connected-devices-worldwide/ (Accessed 11/10/2022)

van der Zee, R. (2016) The 'Airbnb effect': Is it real, and what is it doing to a city like Amsterdam? [online] https://www.theguardian.com/cities/2016/oct/06/the-airbnb-effect-amsterdam-fairbnb-property-prices-communities (Accessed: 15/11/2021)

Wan, E. (2021) Laboring in electronic and digital waste infrastructures: Colonial temporalities of violence in Asia, *International Journal of Communication*, 1, pp. 2631–2651

Wang, Y., Ma, H., Yang, J. and Wang, K. (2017) Industry 4.0: A way from mass customization to mass personalization production, *Advances in Manufacturing*, 5 (4), pp. 311–320

XR Stories (2022) Flood: Combining XR technologies with immersive storytelling [online] https://xrstories.co.uk/project/flood-combining-xr-technologies-with-immersive-storytelling/ (Accessed: 20/04/2022)

13 Resilience in times of crisis: lessons learnt from COVID-19, and the future resilience of businesses and society

Zahra Borghei Ghomi, Layla Branicki, Stephen Brammer and Martina K. Linnenluecke

1. Introduction

'More-severe-than-expected' adverse change is often well beyond the routine emergency management capacities of organisations, both in the public and private sector. Any type of crisis, particularly if it is rapidly evolving and occurring on a global scale, is introducing significant challenges for organisations – the nature of the impact, duration, as well as the best responses and outcomes remain highly uncertain as the crisis unfolds (Linnenluecke, 2020). Unexpected events, such as the COVID-19 pandemic, financial crises, natural disasters, and political instabilities paralyse economies, industries, and organisations; they introduce uncertainty into financial markets, disrupt supply chains, and typically lead to significant impacts on consumer spending. But how is it that some organisations manage to act faster, respond quicker or deploy strategies at the right time that allow them to bounce back despite insufficient information, whereas others do not?

Researchers have attempted to explain the capability of some organisations to respond faster and/or to recover quicker with the concept of organisational 'resilience' (Linnenluecke, 2017). Looking at the rapid changes brought about by COVID-19, it is evident that they profoundly impacted businesses, and the role of business in society (Brammer, Branicki, & Linnenluecke, 2020). It appears that some organisations were 'more resilient' and able to continue to operate despite significant adversity through lockdowns, social distancing guidelines and regulations that prevented public gatherings and wide-spread travel, and allowed residents to only leave their house for essential reasons.

Prior research on organisational resilience has conceptualised resilience in different ways. Some researchers have argued that resilience is a dynamic capability that develops gradually and keeps an organisation functional in times of unexpected events (e.g., Hamel & Välikangas, 2003; Kahn et al., 2018; Sutcliffe & Vogus, 2003). Other studies view resilience as a process that connects a set of adaptive capacities to respond positively to adversity (e.g., Sun, Buys, Wang, & McAuley, 2011) or a process that develops governance strategies over time (e.g., Carmeli & Markman, 2011). Some conceptualisations suggest that resilience is a generalised capacity to learn and adapt (e.g., Wildavsky, 1988), which suggests that a resilient organisation can make meaning in crises and turn adverse conditions into organisational opportunities.

The question of how organisational leaders can best build resilience is still highly contested – resilience is conceptualised differently across various research streams and contexts, thus making it difficult to generate generalisable principles for developing resilience. This chapter examines the 'lessons learnt' from COVID-19, and new insights that have been gleaned into how organisations can build resilience. These include (1) adaptive business models allowing for rapid innovation, (2) resilient supply chains that are less susceptible to disruptions, and (3) a focus on creating a resilient workforce, particularly in sectors such as front-line healthcare and those considered as 'essential services'. The chapter concludes by offering reflections on how organisations but also society can build resilience to future crises and be prepared for future global challenges such as those arising from climate change. The creation of resilient businesses and societies will likely require substantial investment during normal times which might not pay off until a crisis emerges.

2. Adaptive business models

Crises are unexpected, demanding and complex events that require considerable creativeness to be addressed (Tierney, 2003). As was evident from the COVID-19 crisis, many organisations had to rapidly adapt, and diversify and change their business models to handle the challenges. Examples include a rapid shift to an online delivery of services, a rapid transformation to 'contactless' or takeaway deliveries, and a substantial shift to 'work from home'/ telework during the COVID-19 lockdown. Prior research has argued that more resilient organisations can sustain such rapid adaptations more easily due to their increased capacity to adapt their business models, that is, to change their existing strategies, structures, processes and functions as circumstances change (Hamel & Välikangas, 2003; Limnios, Mazzarol, Ghadouani, & Schilizzi, 2014). This increased capacity has, in turn, been attributed to broader access to slack resources and information processing capabilities (Linnenluecke, 2017; Sutcliffe & Vogus, 2003) that allow organisations to rapidly respond to the requirements of changing environment.

Slack resources can include financial, structural, cognitive, behavioural, and/ or relational resources that can help organisations when encountering crises (Gittell, Cameron, Lim, & Rivas, 2006; McDonald, 2003; Richtnér & Löfsten, 2014). Financial slack resources provide companies with much needed financial reserves to endure tough times (Gittell et al., 2006). Structural slack can include areas or sections of the company that are responsible for innovation and long-term, creative or strategic thinking (McDonald, 2003). Cognitive, behavioural and/or relational slack can allow organisational decision-makers to quickly identify and interpret crisis signals, apply critical, creative and flexible approaches, and draw together knowledge, insights and resources from various sources to resolve the challenge (Richtnér & Löfsten, 2014).

While it is a common assumption that slack resources can positively influence adaptability in times of crisis and act as 'buffers' to cushion companies against adverse impacts, it raises questions about the existence of confounding factors – for example, it stands to reason that it is easier for larger companies to accumulate more substantial resources. Importantly, research has found that a viable business model itself depends on the development and preservation of relational reserves (which are often seen as an important element of organisational resilience) over time (Gittell et al., 2006). As was evident from the COVID-19 pandemic, there have been significant external institutional factors impacting organisational resilience. Some sectors – such as the airline industry – would simply not remain viable without government bailouts. Similarly, many small- and medium-sized businesses would not survive without substantial rescue packages. Consequently, organisational resilience has not simply been determined by organisational-level resources and capabilities alone, but is a function of both organisational-level factors and the broader institutional context within which firms operate (Linnenluecke, 2017).

Beyond the traditional conceptualisation of slack resources, the COVID-19 pandemic has pointed to potential new sources of organisational slack: the ability to repurpose existing resources and capabilities and undertake rapid innovation that allows companies to quickly access new opportunities in a dynamic and complex environment (Lengnick-Hall, Beck, & Lengnick-Hall,

2011). Prior research has examined the effects of innovation on resilience. For instance, Teixeira and Werther (2013) outlined three approaches for innovation, including reactive, proactive, and anticipatory. Resilience was thereby directly linked to anticipatory innovation due to an organisation's focus on anticipating future needs and creating an innovation culture, thus anticipating future change. However, COVID-19 has highlighted limitations with conventional approaches to innovation – for instance, innovation in the pharmaceutical industry typically consists of a lengthy process that starts with the identification of potential drug compounds and then moves towards a meticulous refinement and testing process. However, because of COVID-19, the industry quickly shifted to repurposing (i.e., drug repositioning) to explore if existing products had additional therapeutic uses to treat COVID-19 symptoms (Von Krogh, Kucukkeles, & Ben-Menahem, 2020).

More generally, the repurposing of structural and other organisational resources has allowed several organisations to adapt existing organisational structures or processes (often with minimal additional investments) to create new procedures and opportunities in response to crises. Examples are distilleries that started to produce hand sanitisers in response to product shortages during COVID-19 for front-line workers and citizens within their communities (Von Krogh et al., 2020). In turn, the pandemic has also shown that extremely bureaucratic structures can hinder creativity and adaptive changes (Somers, 2009), as experienced by many companies that were struggling to rapidly transition to online and/or remote work arrangements. Important lessons can be gained from these examples for how companies can use this 'repurposing' to increase resilience and face other future challenges.

3.   Resilient workforce

The COVID-19 pandemic has highlighted the importance of resilient employees who can cope with and quickly adapt to rapidly changing demands. Prior research has argued that the presence of resilient individuals is an important element for organisational resilience (Cooke, Cooper, Bartram, Wang, & Mei, 2019; Luthans, Luthans, & Avey, 2013). For instance, prior studies have suggested that the development of a positive organisational response to a crisis requires significant employee engagement to understand the adverse situation and to proceed with reasonable decisions regarding what individual and collective actions to mobilise (Quinn & Worline, 2008). Vogus and Sutcliffe (2007) have suggested that positive behavioural adjustments and mindful actions at the employee level are required, while Weick (1993) has pointed to the importance of sensemaking capabilities. However, it is still unclear to what extent individual resilience behaviour is learnable and developable through social processes and management practices (Cooper, Wang, Bartram, & Cooke, 2019; Mallak, 1998).

A possible avenue for future research is to conduct in-depth interviews with front-line workers and those providing essential services to investigate whether or not there are avenues for employees to learn how to become more resilient. Other researchers (e.g., Luthans, Avey, Avolio, Norman, & Combs, 2006) have suggested using intervention strategies to foster individual resilience, including the identification of setbacks within the work environment, and the identification of potential realistic options to act and to develop psychological capital. Some research using intervention strategies is already emerging. For example, Labrague and de los Santos (2020) demonstrate the application of interventions to keep nursing teams resilient in times of COVID-19 crisis. Interventions included the provision of adequate psychological support services, stress management techniques, and organisational support (including activities directed at improving the emotional, social, physical, professional and financial wellbeing of nurses). Similarly, Albott et al. (2020) explore the concept of 'Battle Buddies' (i.e., a peer support system based on the peer support model developed by the US Army) to deliver a rapidly deployable psychological resilience intervention strategy for front-line workers.

Related to this point, a new line of enquiry has started to examine the development of workforce resilience through work team resilience (e.g., Gucciardi et al., 2018; Stoverink, Kirkman, Mistry, & Rosen, 2020). Work team resilience is thereby based on the capabilities and resources of employees that are drawn together into a team-based construct. Resilient work teams can return to a normal level of operations quickly through the dynamic interactions among individuals and allow interdependent team members to attain goals despite challenging circumstances. Beyond team resilience, COVID-19 has also demonstrated the importance of collective individual mindfulness (e.g., around washing hands, coughing etiquette, touching surfaces, and so on), and we see substantial cultural and institutional differences in enacting these responses which led to substantially different outcomes. For example, Japan never imposed mandatory restrictions, but government warnings alone were enough to prompt many people to practise risk aversion (An & Tang, 2020). Perhaps most importantly, COVID-19 has refocused our definitions of what constitutes 'essential workers'. Many jobs that have proven vital for the overall socio-economic resilience of a country were probably not those that were 'essential' in the pre-pandemic context.

4.   Resilient supply chains

COVID-19 has exposed the vulnerability of international supply chain and logistics networks. While international trade and outsourcing to foreign countries have proven highly efficient economically in the past, many countries have faced challenges with access to supplies due to border closures, reduced air traffic and changed demand patterns. Prior research has already emphasised the importance of resilient and secure supply chain designs to avoid severe and long-term vulnerability to crises. Redundancy, flexibility (e.g., diversified supply chain/transportation and modular designs), agility and innovation are generally regarded as key design attributes for resilient supply networks (Christopher & Peck, 2004; Jüttner & Maklan, 2011; Kwak, Seo, & Mason, 2018; Rajesh, 2017; Rezapour, Farahani, & Pourakbar, 2017), while characteristics such as complexity and tight coupling are regarded as attributes that can make supply chains more disruptive in times of adversity (Craighead, Blackhurst, Rungtusanatham, & Handfield, 2007; Williams, Gruber, Sutcliffe, Shepherd, & Zhao, 2017).

Prior studies have outlined how proactive actions, such as collaborations among supply chain members, connectivity, information sharing, access to timely information, anticipation, preparedness, monitoring, learning and inventory management can enable supply chain resilience (e.g., Brandon-Jones, Squire, Autry, & Petersen, 2014; Colicchia, Creazza, Noè, & Strozzi, 2019; Dubey et al., 2019; Liu & Lee, 2018; Wieland & Wallenburg, 2013). A significant body of research has focused on modelling the impact of these design attributes on supply networks resilience (Kırılmaz & Erol, 2017; Pavlov, Ivanov, Dolgui, & Sokolov, 2018; Pettit, Croxton, & Fiksel, 2019; Pettit, Fiksel, & Croxton, 2010; Ponomarov & Holcomb, 2009; Zhao, Zuo, & Blackhurst, 2019). Some studies pointed to the need to form new creative partnerships within the supply chain to assure the availability of supplies during a crisis (e.g., Durach & Machuca, 2018; Gabler, Richey Jr., & Stewart, 2017; Gölgeci & Kuivalainen, 2020; Scholten, Sharkey Scott, & Fynes, 2019; Voss & Williams, 2013), and pointed to opportunities arising from the application of new technologies (Hardjono, Lipton, & Pentland, 2020).

The dynamic integration of these various attributes and actions was generally seen as positively contributing to enhancing the resilience of international supply chains and logistic networks. Yet while the existing supply chain literature and models have focused on finding optimal combinations across various supply chain attributes (e.g., redundancy/flexibility), many sectors including the airline/travel and hospitality industry were unprepared for the large-scale of existing disruptions. We are therefore likely to see a greater emphasis on creating more local supply chains in a post-pandemic world (McKnight & Linnenluecke, 2020). Future research will be needed to further explore trade-offs between local and internationally oriented supply chains, as well as the resilience of supply chains to much larger international disruptions than previously anticipated.

5. Future Research Directions

Future research can help us to understand how to rebuild from crises – and potentially in better ways. The study of resilience is generally a study of hindsight – meaning that resilience studies often offer a retrospective diagnosis of what went right or wrong. In the past, scientists have issued many credible warnings about the possibility of an infectious disease pandemic, and in hindsight, it becomes evident that these warnings should have been taken more seriously (Dahl, 2020). We are now in a situation where we have received ample warnings from climate scientists about the possibility of future crises resulting from the unmitigated impacts of climate change, and we do have sufficient data to suggest that climate change will pose a significant risk in the future. However, it appears that the key issue is not a lack of warnings, but to develop timely and large-scale responses. Insights from the COVID-19 response show us how to act in resilient ways: the importance of taking early action, the value of clear communication, but also the necessity to implement international measures to avoid cascading impacts in our highly connected world. Future research can build on these insights from COVID-19 to see how early action can be implemented to mitigate future risks, including those arising from climate change.

The recovery of the economy will require resilient businesses. However, and as detailed earlier, organisational resilience is often not determined simply by organisational-level factors alone, but is also a function of the broader institutional context within which firms operate (Linnenluecke, 2017). This is in line with the resilience literature that suggests that organisational resilience is often not determined by organisational resources and capabilities alone (Linnenluecke, 2017). COVID-19 has shown the importance of significant investments into societal resilience, and future research is needed to see how business and societal resilience can be better integrated. Singapore's previous experience with Severe Acute Respiratory Syndrome (SARS) prompted the country to undertake substantial investment in disease outbreak preparation, healthcare infrastructure and a coordinated task force – investments that were perhaps not considered as essential in other countries, but that significantly facilitated Singapore's overall response to the pandemic.

Future research can further tackle the question around optional investments into greater societal resilience by investing into future-proofing our industries and infrastructure more generally. Many COVID-19 recovery programmes are based on the idea of 'building back better' and supporting a green recovery. The main idea behind these initiatives is to use the recovery from the current crisis to foster other desirable outcomes, such as green innovation, a clean energy transition, and investments into emission reduction targets. Importantly, the overall resilience of our society will require an ability to draw together and integrate different initiatives at different levels. However, the substantial fiscal rescue packages have also led to concerns that societal resilience might be eroded in the long run due to rising debt levels and the reduced capacity of governments to undertake future investments. There appears to be a need here to look further into the economic, social, and environmental impacts, as well as trade-offs and opportunities related to the COVID-19 recovery programmes. Future research can use empirical and analytical approaches to assess the impact of these recovery programmes on organisational and societal resilience.

The main challenge with any type of investment into resilience is that the investment does not always pay off in 'normal times', and that its value only becomes visible once a crisis unfolds – thus creating a risk of maladaptive investment and the practical impossibility of adequately addressing all possible future challenges (Allenby & Fink, 2005). Nonetheless, those countries with substantial investments into essential services, healthcare and communication infrastructures are those that are ahead with their response to COVID-19. Case studies of countries that have effectively responded to COVID-19 can provide additional insights into how these investments supported an effective crisis response. During COVID-19, the importance of collective mindful action and the impacts of aggregate individual behaviour changes have become particularly evident. Such mindful collective behaviour change will become equally important for tackling other global challenges, such as climate change. The ability to address such global challenges will ultimately not just require top-down policy measures, but a substantial and collective shift of individual practices towards more sustainable ways of consumption, transportation, energy use and living.

6.   Conclusion

The purpose of this chapter was to review the 'lessons learnt' from COVID-19, offer new insights on how organisations embraced resilient responses in this time, and outline promising avenues for future research. It highlighted the fundamental role of adaptive business models, resilient supply chain, and the creation of a resilient workforce to cope with significant adversity. The development of resilient businesses and societies requires rapid innovation in some global challenges such as COVID-19; however, innovation often involves a lengthy and costly process, which suggests the necessity of timely R&D investments upon receiving credible warnings. While it will be increasingly important for decision-makers to build resilience against a range of future challenges, the creation of resilient businesses and societies will likely require substantial investment during normal times which might not pay off until a crisis emerges.

References

Albott, C. S., Wozniak, J. R., McGlinch, B. P., Wall, M. H., Gold, B. S., & Vinogradov, S. 2020. Battle buddies: Rapid deployment of a psychological resilience intervention for health care workers during the COVID-19 pandemic. *Anesthesia and Analgesia*, 131(1): 43-54.

Allenby, B., & Fink, J. 2005. Toward inherently secure and resilient societies. *Science*, 309(5737): 1034-1036.

An, B. Y., & Tang, S.-Y. 2020. Lessons from COVID-19 responses in East Asia: Institutional infrastructure and enduring policy instruments. *The American Review of Public Administration*, 50(6-7): 790-800.

Brammer, S., Branicki, L., & Linnenluecke, M. K. 2020. COVID-19, societalization, and the future of business in society. *Academy of Management Perspectives*, 34(4): 493-507.

Brandon-Jones, E., Squire, B., Autry, C. W., & Petersen, K. J. 2014. A contingent resource-based perspective of supply chain resilience and robustness. *Journal of Supply Chain Management*, 50(3): 55-73.

Carmeli, A., & Markman, G. D. 2011. Capture, governance, and resilience: Strategy implications from the history of Rome. *Strategic Management Journal*, 32(3): 322-341.

Christopher, M., & Peck, H. 2004. Building the resilient supply chain. *The International Journal of Logistics Management*, 15(2): 1-14.

Colicchia, C., Creazza, A., Noè, C., & Strozzi, F. 2019. Information sharing in supply chains: A review of risks and opportunities using the systematic literature network analysis (SLNA). *Supply Chain Management: An International Journal*, 24(1): 5-21.

Cooke, F. L., Cooper, B., Bartram, T., Wang, J., & Mei, H. 2019. Mapping the relationships between high-performance work systems, employee resilience and engagement: A study of the banking industry in China. *The International Journal of Human Resource Management*, 30(8): 1239-1260.

Cooper, B., Wang, J., Bartram, T., & Cooke, F. L. 2019. Well-being-oriented human resource management practices and employee performance in the Chinese banking sector: The role of social climate and resilience. *Human Resource Management*, 58(1): 85-97.

Craighead, C. W., Blackhurst, J., Rungtusanatham, M. J., & Handfield, R. B. 2007. The severity of supply chain disruptions: Design characteristics and mitigation capabilities. *Decision Sciences*, 38(1): 131-156.

Dahl, E. 2020. Warnings unheeded, again: What the intelligence lessons of 9/11 tell us about the coronavirus today. *Homeland Security Affairs*, 16(7): 1-12.

Dubey, R., Gunasekaran, A., Childe, S. J., Papadopoulos, T., Blome, C., & Luo, Z. 2019. Antecedents of resilient supply chains: An empirical study. *IEEE Transactions on Engineering Management*, 66(1): 8-19.

Durach, C. F., & Machuca, J. A. D. 2018. A matter of perspective – the role of interpersonal relationships in supply chain risk management. *International Journal of Operations & Production Management*, 38(10): 1866-1887.

Gabler, C. B., Richey Jr., R. G., & Stewart, G. T. 2017. Disaster resilience through public–private short-term collaboration. *Journal of Business Logistics*, 38(2): 130-144.

Gittell, J. H., Cameron, K., Lim, S., & Rivas, V. 2006. Relationships, layoffs, and organizational resilience: Airline industry responses to September 11. *The Journal of Applied Behavioral Science*, 42(3): 300-329.

Gölgeci, I., & Kuivalainen, O. 2020. Does social capital matter for supply chain resilience? The role of absorptive capacity and marketing-supply chain management alignment. *Industrial Marketing Management*, 84: 63-74.

Gucciardi, D. F., Crane, M., Ntoumanis, N., Parker, S. K., Thøgersen-Ntoumani, C., Ducker, K. J., Peeling, P., Chapman, M. T., Quested, E., & Temby, P. 2018. The emergence of team resilience: A multilevel conceptual model of facilitating factors. *Journal of Occupational and Organizational Psychology*, 91(4): 729-768.

Hamel, G., & Välikangas, L. 2003. The quest for resilience. *Harvard Business Review*, 81: 52–63.

Hardjono, T., Lipton, A., & Pentland, A. 2020. Toward an interoperability architecture for blockchain autonomous systems. *IEEE Transactions on Engineering Management*, 67(4): 1298-1309.

Jüttner, U., & Maklan, S. 2011. Supply chain resilience in the global financial crisis: An empirical study. *Supply Chain Management: An International Journal*, 16(4): 246-259.

Kahn, W. A., Barton, M. A., Fisher, C. M., Heaphy, E. D., Reid, E. M., & Rouse, E. D. 2018. The geography of strain: Organizational resilience as a function of intergroup relations. *Academy of Management Review*, 43(3): 509-529.

Kırılmaz, O., & Erol, S. 2017. A proactive approach to supply chain risk management: Shifting orders among suppliers to mitigate the supply side risks. *Journal of Purchasing and Supply Management*, 23(1): 54-65.

Kwak, D.-W., Seo, Y.-J., & Mason, R. 2018. Investigating the relationship between supply chain innovation, risk management capabilities and competitive advantage in global supply chains. *International Journal of Operations & Production Management*, 38(1): 2-21.

Labrague, L. J., & de los Santos, J. A. A. 2020. COVID-19 anxiety among front-line nurses: Predictive role of organisational support, personal resilience and social support. *Journal of Nursing Management*, 28(7): 1653-1661.

Lengnick-Hall, C. A., Beck, T. E., & Lengnick-Hall, M. L. 2011. Developing a capacity for organizational resilience through strategic human resource management. *Human Resource Management Review*, 21(3): 243-255.

Limnios, E. A. M., Mazzarol, T., Ghadouani, A., & Schilizzi, G. M. S. 2014. The resilience architecture framework: Four organizational archetypes. *European Management Journal*, 32(1): 104-116.

Linnenluecke, M. K. 2017. Resilience in business and management research: A review of influential publications and a research agenda. *International Journal of Management Reviews*, 19(1): 4-30.

Linnenluecke, M. 2020. Resilience in uncertain times, Blog. https:// corporate -sustainability.org/article/resilience-in-uncertain-times/.

Liu, C.-L., & Lee, M.-Y. 2018. Integration, supply chain resilience, and service performance in third-party logistics providers. *The International Journal of Logistics Management*, 29(1): 5-21.

Luthans, B. C., Luthans, K. W., & Avey, J. B. 2013. Building the leaders of tomorrow: The development of academic psychological capital. *Journal of Leadership & Organizational Studies*, 21(2): 191-199.

Luthans, F., Avey, J. B., Avolio, B. J., Norman, S. M., & Combs, G. M. 2006. Psychological capital development: Toward a micro-intervention. *Journal of Organizational Behavior*, 27(3): 387-393.

Mallak, L. A. 1998. Putting organizational resilience to work. *Industrial Management*, 40(6): 8-13.

McDonald, S. 2003. *Innovation, organisational learning and models of slack*. Paper presented at the Proceedings of the 5th Organizational Learning and Knowledge Conference, Lancaster University.

McKnight, B., & Linnenluecke, M. K. 2020. Businesses step up to make the products we need to get through the coronavirus pandemic. https:// theconversation .com/ businesses -step -up -to -make -the -products -we -need -to -get -through -the -coronavirus-pandemic-134505.

Pavlov, A., Ivanov, D., Dolgui, A., & Sokolov, B. 2018. Hybrid fuzzy-probabilistic approach to supply chain resilience assessment. *IEEE Transactions on Engineering Management*, 65(2): 303-315.

Pettit, T. J., Croxton, K. L., & Fiksel, J. 2019. The evolution of resilience in supply chain management: A retrospective on ensuring supply chain resilience. *Journal of Business Logistics*, 40(1): 56-65.

Pettit, T. J., Fiksel, J., & Croxton, K. L. 2010. Ensuring supply chain resilience: Development of a conceptual framework. *Journal of Business Logistics*, 31(1): 1-21.

Ponomarov, S. Y., & Holcomb, M. C. 2009. Understanding the concept of supply chain resilience. *The International Journal of Logistics Management*, 20(1): 124-143.

Quinn, R. W., & Worline, M. C. 2008. Enabling courageous collective action: Conversations from United Airlines flight 93. *Organization Science*, 19(4): 497-516.

Rajesh, R. 2017. Technological capabilities and supply chain resilience of firms: A relational analysis using Total Interpretive Structural Modeling (TISM). *Technological Forecasting and Social Change*, 118: 161-169.

Rezapour, S., Farahani, R. Z., & Pourakbar, M. 2017. Resilient supply chain network design under competition: A case study. *European Journal of Operational Research*, 259(3): 1017-1035.

Richtnér, A., & Löfsten, H. 2014. Managing in turbulence: How the capacity for resilience influences creativity. *R&D Management*, 44(2): 137-151.

Scholten, K., Sharkey Scott, P., & Fynes, B. 2019. Building routines for non-routine events: Supply chain resilience learning mechanisms and their antecedents. *Supply Chain Management: An International Journal*, 24(3): 430–442.

Somers, S. 2009. Measuring resilience potential: An adaptive strategy for organizational crisis planning. *Journal of Contingencies and Crisis Management*, 17(1): 12–23.

Stoverink, A. C., Kirkman, B. L., Mistry, S., & Rosen, B. 2020. Bouncing back together: Toward a theoretical model of work team resilience. *Academy of Management Review*, 45(2): 395–422.

Sun, J., Buys, N., Wang, X., & McAuley, A. 2011. Using the concept of resilience to explain entrepreneurial success in China. *International Journal of Management & Enterprise Development*, 11(2/3/4): 182–202.

Sutcliffe, K. M., & Vogus, T. J. 2003. Organizing for resilience. In K. Cameron, J. E. Dutton, & R. E. Quinn (Eds), *Positive Organizational Scholarship*: 94–110. San Francisco: Berrett-Koehler.

Teixeira, E. d. O., & Werther, W. B. 2013. Resilience: Continuous renewal of competitive advantages. *Business Horizons*, 56(3): 333–342.

Tierney, K. J. (2003). Conceptualizing and measuring organizational and community resilience: Lessons from the emergency response following the September 11, 2001 attack on the World Trade Center. http://udspace.udel.edu/handle/19716/735.

Vogus, T. J., & Sutcliffe, K. M. 2007. *Organizational resilience: Towards a theory and research agenda.* Paper presented at the 2007 IEEE International Conference on Systems, Man and Cybernetics.

Von Krogh, G., Kucukkeles, B., & Ben-Menahem, S. M. 2020. Lessons in rapid innovation from the COVID-19 pandemic. *MIT Sloan Management Review*, 61(4): 8–10.

Voss, M. D., & Williams, Z. 2013. Public–private partnerships and supply chain security: C-TPAT as an indicator of relational security. *Journal of Business Logistics*, 34(4): 320–334.

Weick, K. E. 1993. The collapse of sensemaking in organizations: The Mann Gulch disaster. *Administrative Science Quarterly*, 38: 628–652.

Wieland, A., & Wallenburg, C. 2013. The influence of relational competencies on supply chain resilience: A relational view. *International Journal of Physical Distribution & Logistics Management*, 43(4): 300–320.

Wildavsky, A. B. 1988. *Searching for Safety.* New Brunswick, NJ: Transaction Books.

Williams, T. A., Gruber, D. A., Sutcliffe, K. M., Shepherd, D. A., & Zhao, E. Y. 2017. Organizational response to adversity: Fusing crisis management and resilience research streams. *Academy of Management Annals*, 11(2): 733–769.

Zhao, K., Zuo, Z., & Blackhurst, J. V. 2019. Modelling supply chain adaptation for disruptions: An empirically grounded complex adaptive systems approach. *Journal of Operations Management*, 65(2): 190–212.

14 The need to align research on economic organisations with degrowth

Ben Robra and Iana Nesterova

1. Introduction

Human civilisation is facing unprecedented changes through human-induced climate change, biodiversity loss, ocean acidification, and other forms of ecological degradation (Kallis, 2018). For an ever-increasing number of scholars, the continued ecological degradation and intensified climate change are the result of the unrestricted pursuit for economic growth in the capitalist system. In other words, the ecological "crisis is not being caused by human beings as such, but by a particular economic system: a system that is predicated on perpetual expansion" (Hickel, 2020, p. 1). Hickel (2020) describes capitalism as an economic system that is based on and requires economic growth (i.e. expansion of economic activity) to function. However, in this chapter capitalism is not only seen as an economic system but much rather at once as a socio-economic as well as political economic system based on and leading to economic growth (Foster et al., 2010; Ruuska, 2019). Beyond economic structures, society's structures in general are likewise based on capitalism's imperative of capital accumulation and growth (Büchs and Koch, 2019).

The acknowledgement that the capitalist system and its drive for accumulation and expansion are the core cause of ecological degradation and climate change has profound implications on how to view economic organisations in connection to sustainability. A perspective on organisations that takes a political economic perspective is required. Ergene et al. (2020) remark that the field of 'business sustainability', as well as business schools in general, focus on the business case for sustainability and win–win perspectives. Such perspectives explicitly ignore the political economic system of capitalism, thus making it impossible to address the root cause of the ecological crisis. Further, the aforementioned perspectives often entail a focus on eco-efficiency (Dyllick and Hockerts, 2002) to create economic, competitive advantages and win–win outcomes (see e.g. Porter and Kramer, 2006). Eco-efficiency means producing with less energy/resources per unit. This highlights an approach to sustainability where being green needs to pay off (Ergene et al., 2020). In other words, this approach to sustainability represents a capitalist motivation to reduce costs to ensure capital accumulation.

Not only is the above capitalist approach to a problem created by the modus operandi of the capitalist system an oxymoron in itself, it also leads to further ecological degradation. A focus on eco-efficiency within the capitalist system leads to the rebound effect and backfire (van Griethuysen, 2010). The rebound effect is the phenomenon of absolute increase in resource/energy use despite eco-efficiency measures to reduce ecological impact (Alcott, 2005). Eco-efficiency measures lead to reduction in cost per unit produced, which in the capitalist system either lead to higher production consumption of the good (direct rebound) or other goods (indirect rebound) (Dietz and O'Neill, 2013; van Griethuysen, 2010). Backfire in this context describes the absolute increase in resource/energy use above the use levels before the introduction of eco-efficiency measures. Without acknowledging these inherent aspects of the capitalist political economy, organisational studies on sustainability will fail to contribute to the needed transformation of society towards sustainability. Economic organisations play an important role in this transformation and their study is vital (Nesterova, 2020; Robra et al., 2020). Ergene et al. (2020) therefore call for a radical and normative approach to organisational research that explicitly acknowledges the political economy of the capitalist system. Normative approach meaning here a value-laden approach that seeks to achieve a societal change.

The degrowth discourse not only acknowledges the capitalist system and its growth imperative as the root cause of climate change and ecological degradation (D'Alisa et al., 2015; Hickel, 2020) but has recently connected this to explicit political economic perspectives (Chertkovskaya et al., 2019). Degrowth aims to reduce the absolute societal levels of resource and energy use to ensure a return to operating within planetary boundaries (Robra and Heikkurinen, 2021; Rockström et al., 2009). However, research on economic organisations in the context of degrowth has been meagre (Nesterova, 2020). Yet degrowth represents a radical and normative discourse that would fit well with the call for a shift in focus by Ergene et al. (2020). This book chapter therefore aims to present the need for a research agenda on organisations aligned with degrowth and suggest several directions for such an agenda.

D'Alisa (2019) advocates the use of Gramsci's (1971) concept of hegemony and counter-hegemony to describe degrowth's ideological opposition to the capitalist system. In line with this, the chapter will describe degrowth as a counter-hegemony and its incompatibility with capitalism in Section 2. This section further translates these insights to the organisational level to highlight the resulting incompatibility of business with degrowth. Following these theoretical insights, Section 3 will present the need for a counter-hegemonic research agenda for organisations and what this research agenda might look like. Further, this section will conclude the chapter.

2. Degrowth and its incompatibility with capitalism

As mentioned in the introduction, degrowth aims to reduce the levels of resources and energy used in society in absolute terms (Hickel, 2020). This translates into degrowth aiming to decrease matter-energy throughput of society (Kallis, 2018; Robra and Heikkurinen, 2021). Matter-energy through-put means a unidirectional flow of energy and resources from the environment through the socio-economic system and back into the environment, which begins ultimately with usage (and thus depletion) of low entropy resources and ends with high entropy wastes, that is, pollution (Daly, 1985). Degrowth often gets misinterpreted as merely aiming to reduce economic growth, but this is not the aim per se. Yet, a reduction in matter-energy throughput will very likely lead to a decrease in economic growth as well as overall economic activity as counted in terms of Gross Domestic Product (GDP) (Kallis, 2018; Latouche, 2009). What is important to note here is that in a societal system which heavily relies on the limitless expansion of economic activity (i.e. capitalism) a reduction or indeed negative economic growth will lead to recession and crisis (Hickel, 2020). In a capitalist system various societal structures and functions such as for example health care and pensions are based on and inter-twined with the continued expansion of capital accumulation and economic activity (Büchs and Koch, 2019). Degrowth therefore aims to transform society and its structures to lose their reliance on economic growth to allow for a voluntary downscaling of production and consumption (Kallis, 2018).

Capitalism aims at continued capital accumulation which leads to and requires constant economic expansion (Marx, 1969 [1867]). Degrowth represents a polar opposite to this ideology (Kallis, 2018). Capitalism and its growth imperative represents the dominant ideology in society (Dale, 2012; D'Alisa and Kallis, 2020). Concepts like infinite economic growth, profit maximisation and capital accumulation are depoliticised and largely unquestioned; turning them into common senses[1] (Buch-Hansen, 2018). Capitalism can thus be described as the current hegemony of society. The term 'hegemony' is used here in the Gramscian sense as an ideology that dominates in society through (largely) non-violent means and structures (Fontana, 2008).[2]

The core drivers of climate change and ecological degradation (i.e. the imperative of endless growth and accumulation) identified by degrowth and other post-growth discourses (see e.g. D'Alisa et al., 2015; Jackson, 2011; Victor, 2008) are largely unquestioned under the capitalist hegemony. It is thus to little surprise that concepts such as green growth, green capitalism, or decoupling are peddled as the solution(s) to the various ecological crises. Yet, these concepts focus on continued economic growth and capital accumulation; the very drivers of the ecological crisis. This is highly problematic as these so-called solutions are unable to address or even acknowledge the root causes driving continued climate change and ecological degradation. Even the United Nation's Sustainable Development Goals (SDGs) arguably aim to make continued growth and accumulation possible instead of addressing sustainability (see Hickel, 2019; Robra and Heikkurinen, 2021).

Degrowth is often criticised for not acknowledging the potential of future technological innovation that supposedly could allow for continued economic growth (see e.g. Schwartzman, 2012). Two interrelated concepts relying on such technology in this context come to mind: green growth and decoupling economic growth from ecological impact. Green growth aims to continue economic growth by focusing economic activity and its constituents (e.g. employment, production, consumption) on seemingly more sustainable options and particularly 'greener' industries and products (Kallis, 2018). What is interesting to note is that degrowth would similarly allow for particular sectors (such as renewable energy production) and consumption of selected goods and services to grow. Further, degrowth scholars tend to emphasise that economic activity in the global South must be allowed to increase to meet needs (see e.g. Hickel, 2020). Yet, degrowth makes clear that this selected and equitable increase in economic activity can only be sustainable if overall global production and consumption (i.e. growth) decreases. Under this assumption economic activity and hence growth is capped, and some sectors (such as fossil fuel production), products and services (such as flying for pleasure) are set to diminish. Green growth on the other hand aims to infinitely continue growth, which includes that of destructive industries, providing some 'greening' occurs (e.g. offsets for flying). What is unfathomable here is that such an intent completely disregards thermodynamic principles. Economic activity necessarily relies on resources and energy that come from the finite resource that is the planet Earth. How a concept can aim at infinite increases on a finite planet (Jackson, 2011) is profound.

Proponents of green growth claim that through technological innovation it is possible to decouple economic growth and economic activities from ecological degradation and thus continue to grow indefinitely (Hickel and Kallis, 2020). Further claims include the possibility to dematerialise the economy through, for example, a focus on services instead of material goods. Dematerialisation completely disregards the material base and energy that services still require, even if not directly visible (Jackson, 2011). For example, the often claimed to be dematerialised service industry heavily relies on digital services which require large amounts of energy, particularly servers. Further, devices to access these services (e.g. computers and smartphones) require material such as rare metals that are mined with severe consequences for workers and the environment alike.

Decoupling, on the other hand, can empirically only be observed in relative terms (Parrique et al., 2019). Relative decoupling reduces ecological impact in relative terms, that is, it makes future increases in growth less impactful but does not address the overall impact (Dietz and O'Neill, 2013). Further, relative decoupling relies heavily on eco-efficiency measures that (as previously described) in a capitalist system lead to rebound effects and backfire, effectively not only doing nothing to decrease the absolute ecological impact but potentially increasing it (van Griethuysen, 2010). It should be noted however, that eco-efficiency measures should not be denounced completely, but only as incapable of achieving sustainability on their own and/or in a capitalist system as overall consumption and production are not addressed (Heikkurinen et al., 2019; Robra et al., 2020). Absolute decoupling, on the other hand (i.e. economic growth without ecological impact), has no empirical foundations[3] (Jackson, 2011). Further, Parrique et al. (2019) argue that the technological innovations required to make absolute decoupling even remotely possible seem highly unlikely. That is, technology that could potentially create negative emissions and impacts such as carbon capture. This does not even take the thermodynamic likely implausibility of such technologies into account.

What is becoming blatantly clear is that concepts such as green growth are not aiming to achieve a sustainable economy and society but rather vehemently trying to 'green' an inherently destructive imperative of the capitalist hegemony. The imperative of economic growth has become such a strong common sense that it seems impossible to imagine a societal system without it. Growth and capitalism are historically claimed to lead to wellbeing, despite increasing well-founded critique of this not being the case (Dale, 2012; Kallis, 2018). Degrowth represents a counter-hegemony to the capitalist hegemony by highlighting that a transformation of society is not only possible but also desirable. Degrowth repoliticises common senses around growth and capital accumulation to imagine a different society not based on and requiring infinite expansion of economic activity (Latouche, 2009). A counter-hegemony should not only critique but practically highlight how things can be done differently and in line with alternative common senses (D'Alisa et al., 2013; Kallis, 2018). There are a breadth of projects, initiatives and practitioners that demonstrate how a different world in line with degrowth could be possible (D'Alisa et al., 2015; Liegey and Nelson, 2020). Some of these degrowth-connected initiatives include concepts such as the occupy movement (see Asara and Muraca, 2015) or the back-to-the-landers (see Calvário and Otero, 2015). Yet, the topic and role of economic organisations in this counter-hegemonic context has been largely and surprisingly untouched.

Organisations and particularly economic organisations are vital to societal transformations (Ergene et al., 2020; Mason, 2015) like the one envisioned by the degrowth discourse. Yet, organisations are under-researched in the context of degrowth (Nesterova, 2020; Robra et al., 2020). However, a few studies on the topic exist, albeit some focus on businesses as economic organisations (see e.g. Khmara and Kronenberg, 2018; Roulet and Bothello, 2020). This is problematic in the context of degrowth. The problem with focusing on businesses as economic organisations in the context of degrowth lies in degrowth counter-hegemony in stark opposition to capitalism. This chapter defines businesses as synonymous with firms and corporations (these are the main types of business). In line with this definition, businesses are capitalist economic organisations that intentionally or unintentionally, but still inevitably, reproduce the capitalist hegemony and its destructive imperatives. Regardless of such a definition, many organisational sustainability studies disregard the political economy of capitalism (Ergene et al., 2020). That is, such studies are unable to address the root causes of the ecological crisis, namely capital accumulation and growth, the imperatives of capitalism.

By studying economic organisations without acknowledging political economic implications means studying these organisations in line with the capitalist hegemony. Many business sustainability studies focus on the business case for sustainability (Dyllick and Hockerts, 2002). These do not only disregard the political economy but actively try to fit sustainability around the business case, an inherently capitalist conceptualisation including the imperative of accumulation. That is, sustainability in forms of 'greening' is patched around the microeconomic concepts that are the root causes of the ecological degradation. It should not be surprising that similarly to foci on green growth and decoupling, these approaches concentrate on eco-efficiency gains while disregarding the rebound effect and backfire (Nesterova and Robra, 2022; Robra et al., 2020). Essentially these approaches try to make sustainability approaches still pay out in capitalist terms (Ergene et al., 2020). In other words, these approaches still cater to the idea of capital accumulation while sustainability is an extra.

Sustainability, as a concept and practice, should not be made to fit around the business case. Acknowledging political economy and severity of ecological and societal degradation means that sustainability should be at the very core of economic organisations on the microeconomic level and at the very core of theorising on this level. The term 'economic organisation' instead of business is used deliberately here. As an inherently capitalist organisation, business is incompatible with sustainability by degrowth's definition. On the other hand, alternative economic organisations can arguably be aligned with degrowth. This is, inevitably, conditional. These organisations must fit the counter-hegemony of degrowth.

Degrowth's counter-hegemony implies a transformation of society, its economy as well as its agents and structures away from the capitalist hegemony. This means that degrowth also implies a complete transformation of economic organisations away from capitalism and its imperatives of accumulation and growth. This includes, for instance, organisations' role in society, their aims and practices. In other words, organisations' nature as economic structures and agents must transform. Instead of the capitalist hegemony, economic organisations need to align with the environment as well as humans and non-humans. Further, a deviation from the imperative of profit maximisation is essential for degrowth (Nesterova, 2020).

Economic organisations operating in line with degrowth need to reduce their matter-energy throughput, collectively and individually (Robra et al., 2020). That is, organisations must aim to reduce not only their own throughput but must help to enable other organisations to do the same. This means re-focusing production for wants to production for need satisfaction. This implies that production needs to go far beyond the notion of eco-efficiency. Robra et al. (2020) argue that eco-efficiency alone cannot lead to sustainability but only if seen as a sub-category of eco-sufficiency. Robra et al. (2020, p. 2) define eco-sufficiency as a concept "that focuses on the overall level of production by emphasising 'enough production' (i.e. sufficient levels). 'Enough' relates in this context to sufficient fulfilment of human needs." However, a focus on eco-sufficiency within a capitalist system or indeed by capitalist organisations seems impossible (Robra et al., 2020). Further, products need to be high quality, durable, local wherever applicable, and produced using methods and technologies which are respectful towards life on Earth, both human and non-human. This is in stark opposition to the notion of planned obsolescence often observed in capitalist production (Dietz and O'Neill, 2013; Illich, 2001; Kallis, 2018).

The capitalist hegemony reduces humans to their labour input into the process of production. Further, capitalist production is carried out for the purpose of making profit, the pursuit of which allows organisations to accumulate capital. Degrowth on the other hand presupposes viewing production as a necessarily social process where people spend their lives, and which affects communities of people and non-humans, that is, humans are not just viewed as an input for profit purposes. Economic organisations compatible with degrowth must fully deviate from the pursuit of profit seeking, making and maximisation. Further, economic organisations must refocus their principles of operation from competition and exploitation towards cooperation and wider belonging in the social and natural worlds (Nesterova, 2020). Without profit creation being the reason for an organisation's existence, possibilities for other forms of organising could unfold that aim to satisfy needs. Degrowth as a counter-hegemony in opposition to capitalism requires economic organisations to also be counter-hegemonic. This means the current research focus on economic organisations in the form of capitalist business and firms is unfit for sustainability in line with degrowth.

3.   The need for a new research agenda

Degrowth has a significant impact on how economies should operate to enable a truly sustainable society. It has direct implications for all socio-economic structures and agents embedded within them (Büchs and Koch, 2019). This naturally includes economic organisations themselves. Considering degrowth, research on economic organisations should radically change. The need for a radically new research agenda on economic organisations arises directly from the incompatibility of degrowth and capitalism. Broadly speaking, a new research agenda on economic organisations must actively acknowledge the political economy of capitalism. Further, the research agenda must include a normative perspective opposing capitalism's destructive force on ecology and society.

Stemming from the incompatibility between degrowth and capitalism, the new research agenda on economic organisations must be based on the premise that degrowth is incompatible with any capitalist economic manifestation on the microeconomic level as well. This includes businesses such as firms and corporations. Hence, this research agenda means decidedly leaving behind any attempts to find a common ground between degrowth and capitalism, its structures and agents. This particularly includes capitalist economic organisations. Economic organisations aligned with the capitalist hegemony reproduce capitalism and its destructive force. That is to say that business whether business-as-usual or business-as-usual-but-greener necessarily reproduces capitalism. Instead, the research agenda must focus on alternative economic organisations (i.e. non-capitalist economic organisations).

Considering the incompatibility of degrowth and capitalism, a radically new research agenda must overcome neoclassical economics, that is, "the economics of capitalism" (Rees, 2019, p. 134). In particular, neoclassical *micro*economics, which reduces organisations to firms and firms to profit-maximising entities. Management and organisational studies traditionally and heavily borrow their assumptions from neoclassical microeconomics (Ergene et al., 2020; Luhmann, 2018). Degrowth, on the contrary, is liberated from the domination of neoclassical economics. It borrows from multiple heterodox schools of economics including Marxist (particularly, eco-Marxist), feminist, ecological economics, and other sciences, traditions and philosophies. This must be reflected in degrowth's research agenda on economic organisations.

Particularly promising is social ecological economics (Spash, 2012), as a branch of ecological economics which can inform theorising and research on economic organisations in relation to degrowth. It brings together an ecological and social critique and assumes a critical realist position in terms of philosophy of science (see Bhaskar, 1998, 1989). This signifies embracing normativity and interdisciplinarity. Social ecological economics openly rejects neoclassical economics' assumptions and allows a more adventurous theorising on socio-economic agents and structures far beyond capitalism. Critical realism does not represent the only research philosophy that can be used in the contexts of this chapter. However, the authors of this chapter have made personally very positive experiences with adaptation of this philosophy. Similarly, the chapter does not seek to point towards one methodology that could be used for future research on economic organisations. Yet, the authors, from personal experience, can recommend qualitative research methods including in-depth interviews and particularly case study research to answer research questions arising from this chapter's influence.

Social ecological economics and degrowth theories can be translated to the microeconomic level to inform what economic organisations may look like and what their role is in achieving a degrowth society. Broadly speaking this implies that organisations are not seen as isolated and profit-maximising entities. Instead, economic organisations should be viewed as embedded within society and the environment (Nesterova, 2020). This embeddedness within society further means that economic organisations must also play an active role in society's transformation towards degrowth. The acknowledgement of capitalism and its destructive role on society and environment (Foster et al., 2010) means that organisations play a role in the transformation away from capitalist structures which also means a deviation from capitalist principles at the organisational level. This will also mean that alternative economic organisations will likely struggle within existing capitalist structures before society's transformation to degrowth (Robra et al., 2020). A research agenda which brings together degrowth and economic organisations must include the following:

1. A focus on alternative types of economic organisations beyond business, firms and corporation. This may include alternatives in terms of ownership and governance. Concrete examples of alternative economic organisations that have previously been connected to degrowth are radical cooperatives (see e.g. Blauwhof, 2012) and commons-based peer production organisations (see e.g. Kostakis et al., 2018; Robra et al., 2020). These alternative forms of economic organisation and production are suitable for degrowth because they do not automatically aim to produce for profits and accumulate capital. However, these types of organisation do not automatically align with degrowth and can also be co-opted for capitalist purposes (Robra et al., 2020). A focus on traditional economic organisations such as businesses as firms is only plausible in the context of transition. That is, a transition away from business or similar to alternative economic organisations (see Nesterova, 2020; Nesterova and Robra, 2022). Further these transitions must focus on a move away from the imperatives of capital accumulation and economic growth.

2. Since economic organisations exist within societal structures, they can and must help transform these structures towards ones compatible with degrowth. Research must investigate how economic organisations can participate in the transformation of society towards a degrowth society, which structures constrain, and which structures empower this transformation. For this, keeping in mind the vision of society that degrowth as a counter-hegemonic discourse aims to achieve, is important.

3. It should be expected that capitalist structures will constrain and actively oppose transformation. Further, alternative economic organisations will need to actively resist co-optation (Robra et al., 2020). Resisting co-optation and the nature of struggle against the capitalist system while being within it represent other vital avenues for research in the proposed research agenda. This means researching how alternative economic organisations can resist

co-optation by the capitalist system while still continuing to reproduce as organisations is vital for degrowth.

4. Since the new research agenda emphasises alternative economic organisations, success measures will have to change. In a capitalist society success is measured in monetary terms (Jackson, 2011). In terms of organisations, these are measures such as profit, turnover, sales, and growth. In a degrowth society, what is meant by success, and hence the measures of success, must change on a societal level as well as the level of organisations. Instead of producing for the imperative of profit maximisation, in a degrowth society economic organisations will have to produce for needs satisfaction (Nesterova, 2020). Hence, success may be measured in terms of, for instance, needs satisfied. How success can be thought of, theorised and measured in relation to organisations is yet another strand the proposed research needs to focus on.

The four points above do not represent an exhaustive list but are much rather a start to actively imagine a different research agenda. Researching the microeconomic level, that is, the level of economic organisations is vital for the degrowth discourse which has traditionally focused on matters on the macroeconomic level. Overlooking the microeconomic level can, and does, result in co-optation of degrowth. Recent publications (see e.g. Khmara and Kronenberg, 2018; Roulet and Bothello, 2020) disregard many of the key points outlined in this chapter and simply maintain neoclassical economic assumptions on the microeconomic level. Researchers looking into economic organisations in relation to degrowth need to reimagine how to conceptualise and research economic organisations instead of falling back on theories and methods that maintain the status quo by reforming capitalism and its mode of production.

The likes of "corporate social responsibility, green accounting, investment in new technologies, sustainable development and [...] a 'Green circular inclusive sustainable smart economy'" (Spash, 2020a, p. 122) are examples which distract and reproduce capitalism and from which research on economic organisations should be emancipated. Researchers trying to create business cases for degrowth (Khmara and Kronenberg, 2018) or advocate degrowth as tools for competitive advantages (Roulet and Bothello, 2020) might not actively aim to co-opt degrowth but do so regardless. That is, this research reproduces the capitalist hegemony instead of opposing it. As mentioned before, businesses might be part of a transition towards a degrowth society, this part is, however, to transform away from business and its capitalist imperatives. In other words, businesses defined as capitalist economic organisations cannot be the agents for transformative change. As long as businesses remain businesses, all they can achieve is reform instead of needed transformation and the reproduction of the capitalist hegemony instead of a counter-hegemony.

Creating business cases or competitive advantages represents a continuation of the status quo that is the capitalist hegemony. Degrowth as a counter-hegemony seeks to transcend and overcome this hegemony. Hence, the idea to make degrowth fit a business context (and therefore a capitalist context) ignores the counter-hegemonic and political economic arguments made in this chapter. It can be compared to the problem of critiquing economic growth without critiquing the capitalist system or hegemony (see Spash, 2020b). This means critiquing a symptom of a bigger societal system and its structures without critiquing the root cause, that is, capitalism. It is important to understand that degrowth as a counter-hegemony seeks to overcome capitalism, its structures, norms, and culture. This means a completely radical societal transformation that cannot be achieved through appealing towards reformist transitions (Nesterova and Robra, 2022).

This chapter calls on junior and senior researchers alike to use the arguments made in this chapter as a start to reimagine research on economic organisations to become truly sustainable by helping achieve a sustainable society. For undergrad and postgrad students, this chapter hopes to have created enough interest to consider going down paths less travelled and consider more heterodox approaches as worthwhile alternatives to standard theses approaches. To conclude, as the chapter's authors we maintain that a truly sustainable society is a society aligned with degrowth. Research on economic organisations must therefore similarly align with degrowth and its principles.

Notes

1. Common senses exist as a plural only in Gramsci's (1971) conceptualisation of the term.
2. Albeit it should be stated that the ideological domination of capitalist ideology was never fully non-violent as the capitalist colonial expansion shows. Generally speaking, the capitalist hegemony has become more violent again in recent decades. Yet, the capitalist hegemony presents itself as non-violent. Due to the colonial history of capitalism, degrowth has also been connected to the post-development discourse (see Escobar, 2015).
3. Studies claiming to show empirical absolute decoupling, do so on national levels and only achieve this through either accounting tricks where ecological impact of imported goods is accounted towards other nations or with very small levels of growth over relatively short time periods.

References

Alcott, B., 2005. Jevons' paradox. *Ecological Economics*, 54, 9–21. https://doi.org/10.1016/j.ecolecon.2005.03.020

Asara, V., Muraca, B., 2015. Indignados (Occupy), in: D'Alisa, G., Demaria, F., Kallis, G. (Eds.), *Degrowth – A Vocabulary for a New Era*. Routledge, New York; London, pp. 169–171.

Bhaskar, R., 1989. *Reclaiming Reality: A Critical Introduction to Contemporary Philosophy*. Verso, London.

Bhaskar, R., 1998. *The Possibility of Naturalism: A Philosophical Critique of the Contemporary Human Sciences*, 3rd ed. Routledge, London.

Blauwhof, F.B., 2012. Overcoming accumulation: Is a capitalist steady-state economy possible? *Ecological Economics*, 84, 254–261. https://doi.org/10.1016/j.ecolecon.2012.03.012

Buch-Hansen, H., 2018. The prerequisites for a degrowth paradigm shift: Insights from critical political economy. *Ecological Economics*, 146, 157–163. https://doi.org/10.1016/j.ecolecon.2017.10.021

Büchs, M., Koch, M., 2019. Challenges for the degrowth transition: The debate about wellbeing. *Futures*, 105, 155–165. https://doi.org/10.1016/j.futures.2018.09.002

Calvário, R., Otero, I., 2015. Back-to-the-landers, in: D'Alisa, G., Demaria, F., Kallis, G. (Eds.), *Degrowth – A Vocabulary for a New Era*. Routledge, New York; London, pp. 143–145.

Chertkovskaya, E., Paulsson, A., Barca, S. (Eds.), 2019. *Towards a Political Economy of Degrowth*. Rowman & Littlefield Publishers, London.

Dale, G., 2012. The growth paradigm: A critique. *International Socialism*, 134, 55–88.

D'Alisa, G., 2019. The state of degrowth, in: Chertkovskaya, E., Paulsson, A., Barca, S. (Eds.), *Towards a Political Economy of Degrowth, Transforming Capitalism*. Rowman and Littlefield International, London; New York, pp. 243–257.

D'Alisa, G., Demaria, F., Cattaneo, C., 2013. Civil and uncivil actors for a degrowth society. *Journal of Civil Society*, 9, 212–224. https://doi.org/10.1080/17448689.2013.788935

D'Alisa, G., Demaria, F., Kallis, G. (Eds.), 2015. *Degrowth – A Vocabulary for a New Era*. Routledge, New York; London.

D'Alisa, G., Kallis, G., 2020. Degrowth and the state. *Ecological Economics*, 169, 106486. https://doi.org/10.1016/j.ecolecon.2019.106486

Daly, H.E., 1985. The circular flow of exchange value and the linear throughput of matter-energy: A case of misplaced concreteness. *Review of Social Economy*, 43, 279–297. https://doi.org/10.1080/00346768500000032

Dietz, R., O'Neill, D., 2013. *Enough is Enough*. Routledge, London.

Dyllick, T., Hockerts, K., 2002. Beyond the business case for corporate sustainability. *Business Strategy and the Environment*, 11, 130–141. https://doi.org/10.1002/bse.323

Ergene, S., Banerjee, S.B., Hoffman, A.J., 2020. (Un)sustainability and organization studies: Towards a radical engagement. *Organization Studies*. https://doi.org/10.1177/0170840620937892

Escobar, A., 2015. Degrowth, postdevelopment, and transitions: A preliminary conversation. *Sustainability Science*, 10, 451–462. https://doi.org/10.1007/s11625-015-0297-5

Fontana, B., 2008. Hegemony and power in Gramsci, in: Howson, R., Smith, K. (Eds.), *Hegemony: Studies in Consensus and Coercion*. Routledge, New York, pp. 80–106.

Foster, J.B., Clark, B., York, R., 2010. *The Ecological Rift: Capitalism's War on the Earth.* Monthly Review Press, New York.

Gramsci, A.F., 1971. *Selections from the Prison Notebooks of Antonio Gramsci.* Lawrence & Wishart, London.

Heikkurinen, P., Young, C.W., Morgan, E., 2019. Business for sustainable change: Extending eco-efficiency and eco-sufficiency strategies to consumers. *Journal of Cleaner Production,* 218, 656–664. https://doi.org/10.1016/j.jclepro.2019.02.053

Hickel, J., 2019. The contradiction of the sustainable development goals: Growth versus ecology on a finite planet. *Sustainable Development,* 27, 873–884. https://doi.org/10.1002/sd.1947

Hickel, J., 2020. What does degrowth mean? A few points of clarification. *Globalizations,* 1–7. https://doi.org/10.1080/14747731.2020.1812222

Hickel, J., Kallis, G., 2020. Is green growth possible? *New Political Economy,* 25, 469–486. https://doi.org/10.1080/13563467.2019.1598964

Illich, I., 2001. *Tools for Conviviality.* Marion Boyars, London.

Jackson, T., 2011. *Prosperity without Growth: Economics for a Finite Planet,* Reprint edition. Routledge, London; Washington, DC.

Kallis, G., 2018. *Degrowth.* Agenda Publishing, Newcastle upon Tyne.

Khmara, Y., Kronenberg, J., 2018. Degrowth in business: An oxymoron or a viable business model for sustainability? *Journal of Cleaner Production,* 177, 721–731. https://doi.org/10.1016/j.jclepro.2017.12.182

Kostakis, V., Latoufis, K., Liarokapis, M., Bauwens, M., 2018. The convergence of digital commons with local manufacturing from a degrowth perspective: Two illustrative cases. *Journal of Cleaner Production,* 197, 1684–1693. https:// doi .org/ 10 .1016/ j .jclepro.2016.09.077

Latouche, S., 2009. *Farewell to Growth.* Polity, London.

Liegey, V., Nelson, A., 2020. *Exploring Degrowth: A Critical Guide.* Pluto Press, London.

Luhmann, N., 2018. *Organization and Decision.* Cambridge University Press, Cambridge, UK; New York.

Marx, K., 1969. *Das Kapital – Kritik der politischen Ökonomie – Erster Band.* Dietz Verlag, Berlin.

Mason, P., 2015. *PostCapitalism: A Guide to Our Future.* Allen Lane, London.

Nesterova, I., 2020. Degrowth business framework: Implications for sustainable development. *Journal of Cleaner Production,* 262. https://doi.org/10.1016/j.jclepro.2020.121382

Nesterova, I., Robra, B., 2022. Business for a strongly sustainable society?, in: D'Amato, D., Toppinen, A., Kozak, R. (Eds.), *The Role of Business in Global Sustainability Transformations.* Routledge, London.

Parrique, T., Barth, J., Briens, F., Kerschner, C., Kraus-Polk, A., Kuokkanen, A., Spangenberg, J.H., 2019. *Decoupling Debunked: Evidence and Arguments Against Green Growth as a Sole Strategy for Sustainability.* European Environmental Bureau, Brussels.

Porter, M.E., Kramer, M.R., 2006. Strategy and society: The link between competitive advantage and corporate social responsibility. *Harvard Business Review.* https://hbr.org/ 2006/ 12/ strategy -and -society -the -link -between -competitive -advantage -and -corporate-social-responsibility (accessed: 11.2.16).

Rees, W.E., 2019. End game: The economy as eco-catastrophe and what needs to change. *Real-World Economic Review,* 87, 132–148.

Robra, B., Heikkurinen, P., 2021. Degrowth and the Sustainable Development Goals, in: Leal Filho, W., Azul, A.M., Brandli, L., Lange Salvia, A., Wall, T. (Eds.), *Decent Work and Economic Growth*. Springer International Publishing, Dordrecht, pp. 253–262. https://doi.org/10.1007/978-3-319-95867-5_37

Robra, B., Heikkurinen, P., Nesterova, I., 2020. Commons-based peer production for degrowth? The case for eco-sufficiency in economic organisations. *Sustainable Futures*, 2, 100035. https://doi.org/10.1016/j.sftr.2020.100035

Rockström, J., Steffen, W., Noone, K., Persson, A., Chapin III, F.S., Lambin, E., Lenton, T., Scheffer, M., Folke, C., Schellnhuber, H.J., others, 2009. Planetary boundaries: Exploring the safe operating space for humanity. *Ecology and Society*, 14, 32.

Roulet, T., Bothello, J., 2020. Why 'de-growth' shouldn't scare businesses. *Harvard Business Review*. https://hbr.org/2020/02/why-de-growth-shouldnt-scare-businesses

Ruuska, T., 2019. *Reproduction Revisited: Capitalism, Higher Education and Ecological Crisis*. MayFlyBooks, Minneapolis.

Schwartzman, D., 2012. A critique of degrowth and its politics. *Capitalism Nature Socialism*, 23, 119–125. https://doi.org/10.1080/10455752.2011.648848

Spash, C.L., 2012. New foundations for ecological economics. *Ecological Economics*, 77, 36–47. https://doi.org/10.1016/j.ecolecon.2012.02.004

Spash, C.L., 2020a. The revolution will not be corporatised! *Environmental Values*, 29, 121–130. https://doi.org/10.3197/096327120X15752810323968

Spash, C.L., 2020b. Apologists for growth: Passive revolutionaries in a passive revolution. *Globalizations*, 18 (7), 1123–1148. https://doi.org/10.1080/14747731.2020.1824864

van Griethuysen, P., 2010. Why are we growth-addicted? The hard way towards degrowth in the involutionary western development path. *Journal of Cleaner Production*, 18, 590–595. https://doi.org/10.1016/j.jclepro.2009.07.006

Victor, P.A., 2008. *Managing without Growth: Slower by Design, Not Disaster*. Edward Elgar Publishing, Cheltenham, UK; Northampton, MA.

Index